TRACTOR
1443
MAINE 1940
FARMALL

FARMALL CENTURY

1923–2023

THE EVOLUTION OF RED TRACTORS AND CRAWLERS IN THE GOLDEN AGE OF INTERNATIONAL HARVESTER

LEE KLANCHER

OCTANE PRESS

Octane Press, Edition 1.0, September 2023

On the cover: Farmall M owned by Timothy Rexford of Loch Sheldrake, New York, photographed in a photo studio built at the Half Century of Progress Show in 2019 in Rantoul, Illinois. *Lee Klancher*

On the front end papers: LEFT: Mule skinner and his team in Walla Walla County, Washington. *Russell Lee, Library of Congress LC-USF34-039862-D*
RIGHT: Tractor-drawn potato digger near Caribou, Maine. Photo dated October 1940. *Jack Delano, Library of Congress LC-USF34-041792-D*

On the rear end papers: LEFT: Farmall on cooperative farm in Box Elder County, Utah. *Russell Lee, Library of Congress LC-USF33-012862-M1*
RIGHT: Fuel station on the Aldridge plantation in Mississippi. Photo dated July 1937. *Russell Lee, Library of Congress LC-USF34-017604C*

On the frontispiece: Mechanic at work in a tractor and combine repair shop in Colfax, Washington. Photograph dated July 1941. *Russell Lee, Library of Congress LC-USF34-070108-D*

On the title page: 1912 International Titan 45. *Lou Buice Collection / Lee Klancher*

On the dedication page: Hot rod Farmall M photographed when owned by the late Jerry Mez of Farmall-Land USA. *Lee Klancher*
On the contents page: 1954 Farmall Super M at work with collector Bill Burnham at the helm. *Lee Klancher*

On the back cover: Farmall F-30 Narrow owned by Andrew Dawson. *Lee Klancher*
Sketches circa 1950, likely done by IH industrial designer T. H. Koeber. *Gregg Montgomery Collection*

ISBN: 978-1-64234-139-3
Anniversary Edition ISBN: 978-1-64234-167-6

Library of Congress Control Number: 2022944532

Edited by Faith Garcia
Designed by Tom Heffron
Proofread by Kari Cornell

OCTANE PRESS
octanepress.com

Octane Press is based in Austin, Texas

Printed in China

DEDICATED TO THE MEMORY OF

GARY KRULTZ AND **JERRY MEZ**

McCORMICK
FARMALL
560
CASE IH

THAT OLD FARMALL MAGIC

Just the name "Farmall" takes me back in time, and I don't think I'm alone in that. For me, there are just so many memories of being with my dad out in the field, sitting on the seat with him as a kid, and flying down the road in road gear. I'll never forget that first time driving it all by myself, when he got off and let me drive across the field. I think so many people who grew up with those machines have one of those stories.

When we did the photoshoot of my Farmall in downtown Chicago, it was about six o'clock on a desolate Sunday morning. There was a guy jogging, and he came running across the Michigan Avenue Bridge yelling, "It's a Farmall! It's a Farmall!" He was a businessman attending a meeting in the city, and he hadn't seen one up close since he'd been a kid in South Carolina. He stayed with us until we loaded it up and went down the street. He was there for forty-five minutes and he wouldn't leave. You can tell it just brought back so much emotion for him. So, the machines are time machines, you know? They put us in touch with our past, bringing back memories of our heritage and our family.

I think the other thing that comes to mind is that so many times when you look at a tractor, there's something that a guy did to it. There's something unique that the farmer or his shop in town would do to make it work better for him. Farmers are such problem solvers, and when a tractor doesn't quite work for them, they have a habit of firing up a welder or bolting on a piece that makes it do what they need it to do.

As you know, in many instances, when you look at that drawbar on an H or an M, it's been reinforced in the middle. Have you noticed that? And it's a guy in the town shop who had to do that, or maybe the farmer did it proactively to make sure that it wasn't going to let him down when he was using it. Sometimes with brackets that have been installed, you'll see something sticking up on the light bar and it just doesn't look right. You're thinking, "okay, why did he put that on there?" Is there something on the platform, maybe a lever sticking out, and you're trying to figure out where this would've fit into everything.

I personally love just looking at these improvements and trying to determine what was going on. How did he drill that pattern of holes? Why did that pattern wind up on that frame? What was this guy doing anyway? And I guess that's the thing that comes to mind. There were so many of those. They didn't take it to the degree of Elmo Meiners, but there were many, many farmers who took that little step to make that thing their tractor.

Farmall tractors are time machines, whether they take you back to your own past or prompt you to think about the challenges someone else might have faced. I see these machines so often, and they take me on a little trip through time. I love that.

Max Armstrong was a business and ag-business reporter at WGN Radio in Chicago for more than thirty years, as well as the host of the syndicated programs Farm Progress America *and* Max Armstrong's Midwest Digest*, and co-host of the television show* This Week in AgriBusiness*. He is the author of the book,* Stories from the Heartland.

MAX AND HIS 560

Max Armstrong and his 1962 Farmall 560, photographed at the 2019 Half Century of Progress Show. *Lee Klancher*

FARMALL
DIESEL
McCORMICK
INTERNATIONAL
DIESEL
7up
Friendly

THE GEARHEAD GENIUSES WHO BUILT FARMALL

I grew up in farm country, where pretty much everyone I knew could get their hands dirty, and every block had someone with a mechanical mind who could fix just about anything. If my nose wasn't buried in a book, I was tearing up the county roads and trails with dirt bikes, snow machines, and pickup trucks. When I broke things—which was often—I knew which neighbors had engine hoists, welders, or whatever tool I lacked.

I was taken back to those days during a photoshoot for this book. I had spent the afternoon making photos with Ryan Peters, a collector from Iowa. His garage was full of cruiser motorcycles, modern snowmobiles, and vintage tractors. Talking gearhead with him took me back to my teen years in Wisconsin.

While standing outside his shop late that night, Ryan told me about his friend, a tractor puller who was a genius at hot-rodding Farmall 460 engines. We were talking shaved heads and hand-formed cam lobes, and how the guy deliberately kept his 460 as ratty as possible so when he showed up to tractor pulls everyone assumed it was junk that would hardly get off the blocks.

What hit me that night were the parallels between the ability to turn a wrench, the genius to tune a motor, and the innovators who built the Farmall and all the other mechanical inventions that shaped this crazy modern world in which we reside. The commonality was that all these people lived by the credo that if you need something better, you make it. When rural kids like me broke a drive chain trying to jump the gravel pit road, we had to cobble together master links or fashion a pin from a nail because the shop was thirteen miles away, didn't stock much anyway, and was closed on weekends.

Take that another step, let's say to a rancher in Montana. If her tractor breaks, the same applies, except she's not chasing neighborhood glory, she's losing money and crops if she doesn't quickly solve the problem with a bolt, welder, or whatever happens to fall to hand.

That grit and self-determination is part of being a farmer and part of being an American. I think it was particularly essential in the days when the country was being settled. The folks that immigrated here left behind an entire world to live in one where they could build their own fortune.

The gearheads of those generations—people like Ed Johnston, Bert Benjamin, Alex Legge, and John Steward—all were born with the natural inclination to understand mechanical creations and the temerity and intelligence to envision how those mechanical creations could change the world.

Put these high-horsepower gearhead brains into the fertile petri dish for innovation that was America at the turn of the twentieth century, and the result was a technological revolution that changed the world from a place where most people devoted their time gathering food to one where almost all of our food is gathered in by a hardworking few.

That gearhead innovation built the original Farmall, and that is the story we are dealing with here today. Roll up and dig in—let's go back in time and visit with the hot rod geniuses who transformed our world with agricultural equipment.

Lee Klancher

June 22, 2023

GARAGE MAGIC

This old garage is next to the former museum, Farmall-Land, in Avoca, Iowa. The late Jerry Mez and photographer Lee Klancher had to push out an Oldsmobile muscle car to put Jerry's DLD-2 in place for this photo. Jerry is shown in the back, gamely pretending to spin wrenches. *Lee Klancher*

THE SMITH

Library of Congress LC-USZ62-101317

CHAPTER ONE

THE FOUNDATION

THE ROOTS OF IHC ARE PLANTED
1850–1902

"Mechanics became millionaires,
and millionaires became mechanics."
—Herbert N. Casson, *The Romance of the Reaper*

US National Archives

When Alexander Legge was late coming home from school, his father, Sandy Legge, knew precisely where to find him. The boy's walk home through the community of Jimtown passed a blacksmith shop, and he would often stop to examine the farm machines being repaired there.

Sandy Legge had immigrated from Scotland in 1857, following his sister to Wisconsin, confident that his skills as a veterinarian and his boundless energy would serve him well in the growing nation.

Located midway between Verona and Paoli, in Dane County, the tiny community had a trading store and a Scottish Presbyterian church as well as the blacksmith shop favored by the lad. Sandy moved his family to a log cabin not far from nearby Mt. Vernon and established a small veterinarian hospital. The family struggled to make ends meet, with Sandy forced to augment his farm income by breaking rock and trading horses to scrape up a bare existence.

Harvest time was particularly difficult for the Legge family. Even with wheat prices strong and improving, finding enough help to cut the crop by hand was never easy, particularly when men were drawn away to fight in the Civil War.

Born in 1866, Alex Legge grew up chasing cinch bugs and waterfowl from his father's fields and teaching his pet hunting cow to hide him so he could sneak up on the ducks and geese. His father's philosophy of farm management was that everything on the place was for sale, and he proved it by selling Legge's prized stalking cow for a handsome profit.

The boy complained, to which his father replied, "Train anither. It's good business, lad!"

His father's comment reflected not only the tough pragmatism shared by many of the immigrants who came to America in the nineteenth century, but also the fertile environment in which they thrived.

The country was wild, and enterprising pioneers from around the globe were drawn to a place where they could work their way into a better situation. In the case of the Legge family, his talent training cows was quickly recognized as an opportunity to advance.

This ethic permeated early American society and rewarded the toughest of the early pioneers who built farms, small businesses, and towns. Those who possessed the magical combination of drive, ambition, and intellect—or possibly the guile, greed, and salesmanship—flourished and created everything from the plow and the reaper to steamboats and railroads.

After tough harvests in 1874 and 1875, and after losing roughly $8,000 in money loaned to a friend, his father followed the growing trend of westward expansion and accepted an offer to partner in a 2,000-acre ranch in Colfax County, Nebraska.

Ten-year-old Alex took to his new environment with enthusiasm, fascinated with how many cattle would be on the land. How would they farm such a vast expanse? What would his role be in the operation?

Legge's curiousity would bloom in the new ranch. His father began calling Legge to his bedside each night, where the two of them would puzzle over the issues of the day on the ranch. While Legge did not get much formal schooling in this portion of his life, these consults would serve as early training for Legge to understand how to manage and grow a large wheat operation—skills that he would keep for life.

He saw vast teams of horses laboring for weeks on end to cut the thick grass into fields. He also put in hours of hard labor seeding and harvesting the more than 1,500 acres of small grain grown on the farm—all by hand.

Legge was trusted with buying and selling supplies and livestock and, in his teens, took over the decisions about selecting and purchasing farm machinery. "Machinery interested Legge more than livestock. He spent every minute studying it. As it came into use later, he had an almost uncanny understanding of its mechanism. If there was trouble with any machine, he was summoned. By virtue of his natural mechanical gifts, he became the official repair man of the ranch."[1]

By the time he was fifteen years old, he was the man making the calls on the growing operation while his father was away. "If anything oot o' the ordinar turns up, ask Sandy [Alex's nickname]. He'll ken."[2] The farm hands took orders from Legge easily—a natural sense for working with others was a talent of his that showed up early and became critical later in his career. Also at age fifteen, Legge and his brother, James, were sent to the nearby Brewery Hill School for two terms to study algebra, history, grammar, and composition. Legge did well in all his classes and showed great talent for mathematics.

After nearly being frozen to death in the historic blizzard of 1893, Alex Legge was diagnosed with a heart condition. The doctor recommended a drier climate, and the young man was sent to work on the V-R Ranch in Wyoming. Along with his friend, George Wertz, Legge proved such an able fence rider that he was promoted to serve as a courier, transporting sums of cash through remote and at times lawless territory.

When his mother became ill, Legge applied for a job as a collections agent for McCormick in Omaha. His mother's health was waning, and this would allow him to be close enough for frequent visits.

As the twentieth century loomed, collecting debts was a significant part of the farm equipment business. Machinery was sold to farmers on terms of credit, and company salespeople were more eager to close a deal than to investigate the creditworthiness of the buyer.

Farm crop prices, even today, fluctuate with the weather, market conditions, and government policies. The environment

SANDY LEGGE (1868)

Alex's father immigrated from Scotland with no formal education. A veterinarian, farmer, and savvy manager, he came to the US with nothing and worked his way up to owning a successful 2,000-acre ranch in Nebraska. *From* Alexander Legge 1866–1933

THE LEGGE CHILDREN (1879)

James, George, and Alex (on right) Legge. Alex was thirteen years old when this was taken and was already taking on management responsibilities at the family farm. *From* Alexander Legge 1866–1933 was particularly volatile in the late 1800s. As the world transitioned to a truly global market, prices fluctuated to a larger degree than ever before. Farmers were also required to produce more to make a profit, which meant a larger crop and greater risk. To take those risks, they had to purchase expensive equipment and rely on creditors—which a self-sufficient farmer of previous years may not have had to do.

Farmers who were not sharp at managing the nickel, had bad timing, or just plain bad luck found themselves too deep in debt to recover, and farms were often repossessed by banks.

This led to great unrest on the farm. Organizations formed to protest conditions and large demonstrations took place, some breaking down into riots.[3]

Collecting debts from farmers during this time was no easy task, especially since they had varying degrees of interest in paying their bills. Legge excelled at this work, using logic, charm, and guile to persuade people to pay what they owed. He became the collections manager in Nebraska in 1894, a year marked by regional drought and locusts as well as nationwide political and economic chaos.

In 1896, a twenty-three-year-old son of industry married a daughter of John D. Rockefeller Sr. The newly married man was sent to work at McCormick in Omaha, and the "cowboy bachelor," as Legge was called, was assigned to look after the young man.

The young man was Harold F. McCormick, son of Cyrus H. McCormick, and the assignment would be the biggest break in Legge's career, one that would give him opportunities far beyond hustling cows and talking farmers into paying their bills.

CATTLE DRIVE

Near the Z/T Ranch near Pitchfork, Wyoming, on January 16, 1923. *Charles Belden, Library of Congress LC-USZ62-107981*

THE McCORMICK LEGACY AND MYTH

The McCormick Harvesting Machine Company was founded by Cyrus Hall McCormick Sr., an American businessman who began building mechanical reapers in Virginia and created one of the largest agricultural companies in America.

Born on February 15, 1809, Cyrus Sr. grew up on a farm northeast of Lexington, Virginia, in the heart of the Shenandoah Valley. His Scotch-Irish ancestors arrived in America in 1735, settling first to farm in Pennsylvania and moving to Virginia in 1792.

As a boy, Cyrus Sr. was reserved and didn't make a lot of friends. He preferred to spend time working in his father's carpentry shop rather than out in the fields. He was an excellent horse rider, one of the best in the region.[4]

He grew up in a staunchly Calvinist home, and the Protestant church would be an important part of his entire life. "As a good Calvinist, he found true harmony to exist between success in business and the tenets of his faith."[5] Cyrus Sr. did not use tobacco or alcohol and dressed more formally than his peers. Locals would later recall they thought he was conceited, but this could simply be the interpretation of his stand-off nature, sartorial taste, and strict adherence to his morals.

He survived yellow fever as a small child, possibly because his father, Robert McCormick, refused to administer the doctor's recommendation of heavy bloodletting and instead bathed the boy in bitter herbs and whiskey.

Robert's wise choices were not limited to doctoring. An inventor and first-rate mechanic, he created working examples of a reaper, blacksmith bellows, grist mill, and other farm implements, many of which he patented. The McCormick farm was quite a successful operation and, by 1830, covered approximately 1,200 acres (and can be visited

1831 McCORMICK REAPER

The mechanical reaper was a key agricultural advance of the nineteenth century, allowing the farmer to convert from cutting grain by hand to cutting grain with a horse-drawn machine. This image is from the film *Romance of the Reaper* (1929), produced by the International Harvester Company to celebrate one hundred years of the reaper. *Wisconsin Historical Society / 24863*

today as a historic site). Robert also owned a number of slaves, one of whom became close to Cyrus Sr.

Born in 1808, a year before Cyrus Sr., Jo Anderson became a constant companion, walking Cyrus Sr. to school each day despite the fact he, as a black man, was not allowed to attend. "So we nursed together and played together all over the farm when we were both children," Anderson said. "Massa Cyrus was always quiet and had no wild capers like with some boys . . . [he] was a most elegant rider, no better in all the country."[6]

A skilled blacksmith, Anderson assisted Cyrus Sr. with building his inventions, including the reaper and their iron furnace. Anderson came to Washington, DC, to be with

CYRUS McCORMICK PAINTING

The 1831 test of the McCormick reaper has been mythologized with paintings, drawings, books, movies, and recreations, including this painting done in the 1930s. The original reaper was destroyed when the Chicago fire consumed the McCormick factory. *Bernarda Bryson Shahn / Alamy*

Cyrus Sr. for a time in the 1850s, and Cyrus Sr. offered to bring Anderson to Chicago later in life, where he could be free. Anderson opted to remain in Virginia, and became free after the Civil War ended in 1865. Anderson and Cyrus Sr. exchanged gifts and letters throughout their lives.[7]

In the 1820s, Robert and Cyrus Sr. began working together to develop their farm equipment and business. The two clearly had ambitions beyond just running their farm. They would pursue attempts to build and sell a device that reduced hand labor when harvesting hemp fiber, start and run a smelting business, and they would travel together in June 1831 to file for a patent on a hillside plow Robert designed.

The legend of Cyrus McCormick begins shortly after that trip—as do the events that would become the hotly-contested subject of court cases, newspaper articles, books, and family disagreements.

The McCormick legend begins with a reaper built by Robert in 1816. By all accounts, he abandoned it for some time but returned to it briefly in late spring 1831, where it worked reasonably in straight grain but poorly in tangled wheat.

Sometime between May and July 1831, Cyrus Sr. is said to have developed the machine so that it would cut tangled wheat properly. The work was done with the help of Anderson, who was a fine blacksmith.[8] The key innovation was adding a reciprocating knife, which Cyrus Sr. is said to have built and refined. He said he then constructed a wooden reaper that was an evolution of his father's design.[9] No documentation survived, so what is known of this time is based on conflicting personal recollections.

Coincidentally, Cyrus Sr. and Robert had been in Washington, DC, to patent their hillside plow in June 1831. Six weeks prior to their visit, William Manning patented a reaper that used a vibrating knife to cut grain. There is no evidence the McCormicks encountered Manning's machine while in the office, but if they had seen the moving cutting bar on the reaper model, they would almost certainly have recognized how the knife could improve their design.

The story goes that Cyrus Sr. demonstrated his working reaper at John Steele's farm in late July 1831. According to Cyrus Sr. himself, only three or four people (outside of the

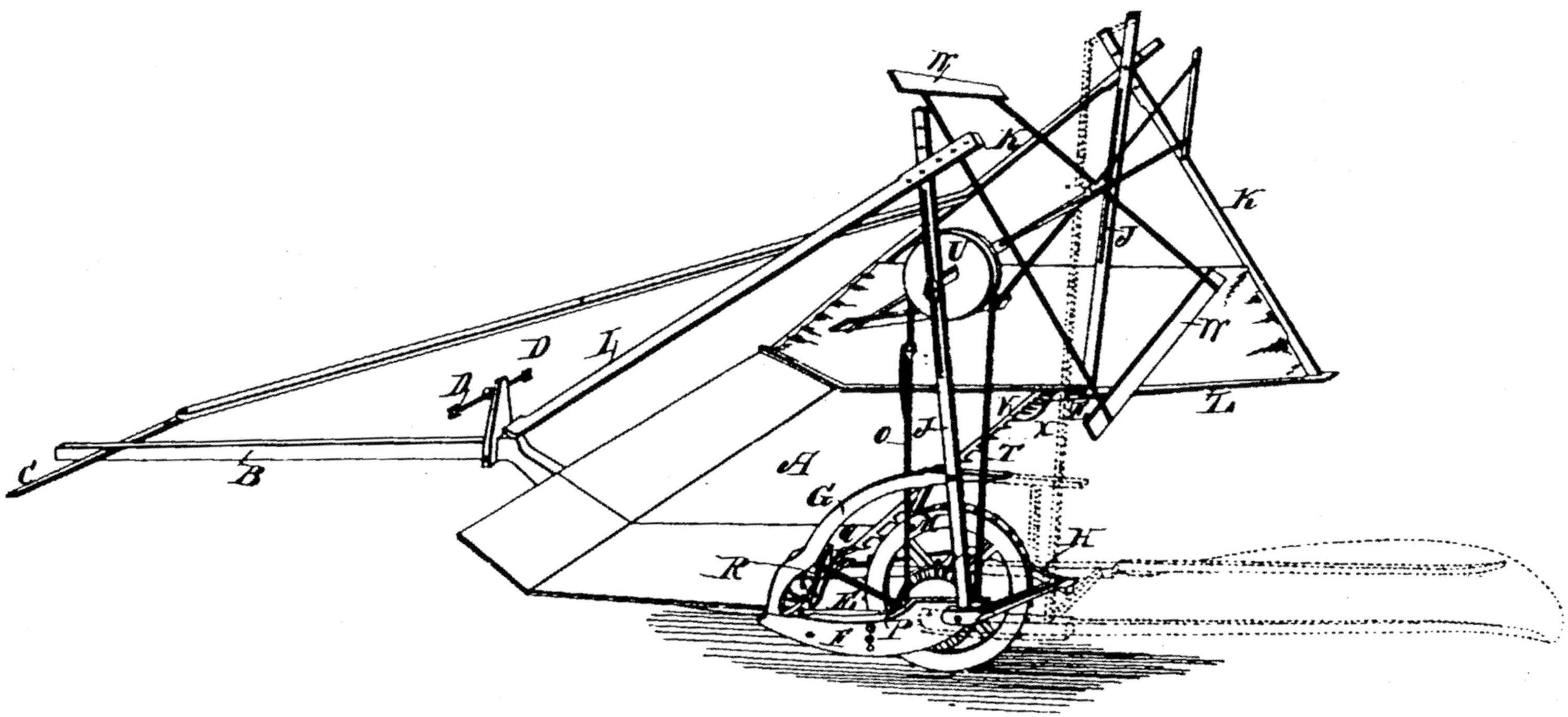

McCORMICK REAPER PATENT

Cyrus H. McCormick's first patent for a reaper was filed in 1834. Obed Hussey filed his reaper patent in 1833. *US Patent Office* family) attended the demonstration, and the machine did not cut the grain perfectly.[10] As depicted in some of the paintings of the event, some recalled Anderson walked next to the machine, while Cyrus Sr. drove the horse. Anderson did not recall the demonstration late in his life. "I don't remember the trail at Steele's place, nor do I remember leading the horses as Polly Carsons say, but it mos' likely I did."[11]

The earliest known documentation of this machine is a letter from March 1832. Cyrus Sr. was writing to his father for advice on developing the reaper in Kentucky, and Robert responded that this would be difficult, as the machine required considerable development that would be better done in Virginia, "free from the watchful and jealous eye of strangers."[12]

ROBERT McCORMICK (1780–1846)

The father of Cyrus Hall McCormick started building a reaper in the early 1800s. Robert's descendants would later claim the reaper was his invention and it was given to Cyrus as a gift. *Wisconsin Historical Society / 8395*

Legends are something the McCormicks found themselves to be nearly as good at manufacturing as machinery, and Cyrus Sr.'s descendants did a fine job of transforming this demonstration into the moment the reaper first appeared to grace farmers—and the world—with its surely God-given presence.

The notion that Cyrus Sr. is the sole inventor of the mechanical reaper is overstated, and as time passed, this "sole inventor" myth became a fabrication the family used to promote their leadership in the field. His contributions to reaper development were important, and he was one of the most successful entrepreneurs of the nineteenth century, but the reaper was developed with significant mechanical contributions by many different inventors over multiple decades.

Cyrus Sr. applied for a very broad patent for a mechanical reaper on June 21, 1834. The story goes that he spotted a photo of Hussey's reaper patent in the April 1834 *Mechanic's Machine* magazine. When he saw that there was competition in the field, he immediately filed a patent of his own.[13]

Savvy at the power of the press from a young age, Cyrus Sr. wrote a letter to *Mechanic's Machine* challenging Hussey to a public competition between their two reapers. Hussey did not have a machine available, and the challenge went unanswered. Their rivalry was only just getting started.

Cyrus Sr. continued to develop his machine, adding metal fingers in front of the cutting bar, changing gearing, and making the cutting height adjustable.[14] He put on a number of demonstrations of his reaper with varying success over the ensuing years, including one where farmer John Ruff, dismayed at the damage the reaper was doing to his crop, tossed Cyrus Sr. and the machine off his land. Other demonstrations went better, with farmers purchasing machines on occasion.

The reaper was a side project at the time, as the McCormicks occupied most of their energy building their smelting business from 1835 to 1839.[15] Cyrus Sr. was also deeded 400 acres by his father in 1835, and transforming that property into a workable farm required considerable time and effort.[16]

In 1839, with the smelting venture failing, the McCormicks turned back to the reaper. Cyrus Sr. began selling machines in 1842 and sold a whopping seven units. He sold a few more in 1843 and 1844—all built in his father's shop on their 532-acre Walnut Grove farm in Virginia.

These meager sales numbers would not last long. While the McCormick family gift for building legend was remarkable, they were even more talented at growing their business.

McCORMICK EXPANDS

The McCormick reaper business expanded steadily, and Cyrus Sr. ran it along with his brothers, Leander James McCormick and William Sanderson McCormick. They licensed at least one maker, Seymour and Morgan, to build machines, and explored a similar arrangement in Cincinnati.[17] This allowed them to sell 123 machines in 1845—Cyrus Sr. would later claim the profit per unit was $20 (adjusting for inflation, this would be $787.29 per unit in 2023).

In 1847, just after Robert McCormick passed away, the family made the wise decision to relocate the company from rural Virginia to Chicago. At the time Chicago had around 17,000 residents, and cities like Milwaukee and Cleveland had better infrastructure, roads (Chicago had very few paved streets), and living conditions. Chicago water was a menace, carrying smallpox and cholera between 1847 and 1850. "[Chicago] had little to offer in the way of material comfort, but it was a city of opportunity."[18]

The move improved access to shipping, as Chicago had unparalleled water transportation access, and it also put the McCormick company square in the middle of a global hotspot for agricultural machine innovation. The bulk of the large agricultural companies were based in the center of the country, and America's fertile ground proved the ideal place for innovators to bootstrap themselves by helping farmers harvest more with less. They initially partnered with Charles M. Gray and Seth Warner of Chicago[19] to manufacture the reapers—arrangements that would not prove satisfactory. In

CYRUS HALL McCORMICK (1809–1884)
A gifted and innovative business manager with a flair for sales and promotion, Cyrus Sr. founded the McCormick Reaper Company. With the help of his brothers, he built it to be the nineteenth century's largest manufacturer of farm machinery. *Pictorial Press Ltd / Alamy*

McCORMICK FARM
This illustration depicts Walnut Grove, the Virginia homestead of the McCormick family. *Wisconsin Historical Society / 63082*

McCORMICK FACTORY 1847

The factory was located on the north bank of the Chicago River, east of the Rush Street bridge. Built in 1847 by Cyrus McCormick and his partner Charles M. Gray, it was destroyed in the Chicago Fire of October 8–9, 1871. *From* Century of the Reaper

1848, Cyrus took on new partners—William B. Ogden and William E. Jones[20]—and the new firm of McCormick, Ogden, & Co. was formed. McCormick would use his profits from the season to buy out his partners by the end of 1849, founding C. H. McCormick & Co. Leander came to Chicago to help him build a new factory, and William came on to help run the finances.

The new company took up shop in a small factory on the north side of the river. The tidy brick building housed thirty-three workers—ten of them blacksmiths—and a variety of equipment powered by a steam engine, including fifteen machine tools.[21]

Production grew explosively. By 1850, more than 120 men worked in the main factory. One year later, Chicago papers would boast the factory was the largest of its kind anywhere in the world.[22] By 1856, McCormick Works was building four thousand reapers a year, and by 1858, the McCormick Harvesting Machine Company was the largest farm equipment manufacturer in the United States.[23]

The speed of the transition was astonishing. In little more than a decade, Cyrus Sr. had transformed his business from a handful of hand-built reapers in his father's shop to an industry leader and one of the largest companies in the nation.

CHICAGO'S EARLY DAYS

When the McCormicks moved their business to Chicago, it was a rough-and-tumble town with cheap real estate and little development. The family wisely invested heavily in property—William even tried to sell their Virginia home to raise money to purchase more Chicago homes, but his fondness for it led him to abandon this plan in 1859. *Library of Congress LC-USZC2-2092*

OBED HUSSEY

The Hussey mechanical reaper, developed by Obed Hussey, was patented in 1833—one year prior to the McCormick mechanical reaper. *Wisconsin Historical Society / 24855*

THE REAPER WARS

The race to design, perfect, build, and sell the mechanical reaper to the farmer in the latter half of the nineteenth century was an epic battle. The machine was a fantastic improvement to standard farming practices, eliminating the need for massive harvest crews, and the technology was rapidly evolving.

Much like tractors in the 1940s and personal computers in the 1980s, mechanical reapers experienced a period of incredible growth in demand and sales. The reaper's long development cycle meant this demand was seen in a series of waves that took place over five decades, and the war for farm consumers spawned an almost unparalleled spate of unsavory behavior that included false advertising, sabotage, illegal contracts, patent law shenanigans, bribing of politicians, and—when all else failed—hand-to-hand combat in the field.

As the machines became more prominent in changing the agricultural landscape, the contest to win customers became more public and heated. Local fields staged high-profile reaper demonstrations. International contests for the new technology became extremely important for sales, and no quarter of expense (or lubrication of contest judges) was spared in pursuit of a victory. Lawsuits also flew fast and furious, as American patent law evolved and competitors turned to the courts in an attempt to protect their market share.

One of the most interesting examples is the case of John Henry Manny, whose reaper drew McCormick's attention when it won several awards at the Paris Exposition in 1855, one of the world's most famous showcases for new technology.[24]

TRUMPETING FIELD TRIUMPHS

This advertisement focuses on winning the field. *Wisconsin Historical Society / 86308*

While the McCormick reaper won top honors at the show, Cyrus Sr. was infuriated by what he believed was a copy of his design and he sued Manny & Co. for patent infringement, asking for $400,000 in damages.[25] Manny hired lawyer Edwin Stanton, and McCormick brought in former US Attorney General Reverdy Johnson. The Manny team tapped a local Illinois lawyer, Abraham Lincoln (yes, that one) to assist in their defense. The case was moved to Cincinnati, so Lincoln did not contribute to the trial (although he was never officially dismissed). Lincoln would, however, appoint Stanton to be his Secretary of War (despite the fact that Stanton objected heartily to Lincoln's presence—legend has it Stanton compared him unfavorably to a primate).[26]

Fascinating historical footnotes aside, McCormick lost to Manny in a lower court. Manny died from a stomach illness in 1856, and, after his death, the McCormicks pushed their case to the Ohio Supreme Court, which also determined there was in fact no patent infringement and the Manny machine was an improvement upon the McCormick patents. This dramatic conflict is cited by the community of Rockford, Illinois—where Manny built his reaper—as a catalyst for the town's nickname as "The Reaper City."[27]

A few years later, in 1861, McCormick's patent renewal was denied in the courts, while competitor Obed Hussey's was renewed—another legal argument directly opposed to the McCormick family claim to be the sole inventor of the reaper.

A sizable list of books were published that explained exactly how and why Cyrus *did not*, in fact, invent the reaper. The list of authors and outspoken (and often litigious) public opponents to the story includes Obed Hussey—a rival inventor—as well as John F. Steward and no fewer than ten[28] members of the McCormick family, including Robert, Cyrus Sr.'s father.

Public drama such as this high-profile disagreement would haunt the McCormicks for nearly a century—much of it replete with lurid headlines and bizarre storylines.

Family squabbles and legends aside, there is a large body of work to peruse, and conclusively stating who invented the reaper is beyond the scope of this book. A reading of all these works leads one to conclude that Cyrus Sr.'s claim to be the sole inventor of the reaper is an exaggeration at best. It is clear that Obed Hussey and others had patents that predated

GATHERING THE FARMERS

This Deering advertisement captures the competitive spirit of the day as well as the widespread notion among capitalists that their success benefited many and was, in fact, God-given. Cyrus Sr., William Deering, and John D. Rockefeller Sr. were extremely pious men who expressed the belief that their wealth was generated due to divine will. *Wisconsin Historical Society / 86832*

the McCormick patent, and there were several court cases that the McCormicks lost based solely on that.[29]

Charles W. Marsh wrote: "I know that the public has fallen into the habit of giving the elder C. H. McCormick full credit for the invention of the modern harvesting machine as if it were his production or a simple improvement on his reaper. As well might it be said that Franklin was the inventor of the telegraph or Morse was the inventor of the telephone. I have no disposition to question his right to his reaper or the greatness of his work; but there is nothing of his, as original invention, in the modern self-binding grain harvester."[30]

The reaper was developed due to multiple innovations by a variety of people from Europe and the United States, including John Henry Manny, Obed Hussey, Charles W. Marsh, Patrick Bell, and others. The result was a machine that transformed American farm life.

JOHN F. STEWARD: THE FIERY GENIUS

As the reaper wars were beginning to erupt on American farms, fourteen-year-old John Fletcher Steward had transformed his natural gift for mechanical things into establishing himself as the "Mr. Fixit" in his childhood home of Plano, Illinois.[31] Slightly prior to 1842, his family had traveled by covered wagon from Wayne County, Pennsylvania, to the fertile fields of Illinois.[32]

The influential journalist Horace Greeley—the editor of the *New York Tribune* who wrote popular articles urging early Americans to venture west—was inspired by the picturesque dells and streams in and around Plano. Greeley's 1871 letter to a young correspondent is credited with coining the line, "Go West, young man." (Scholars differ on the opinion of where that phrase arose, so Greeley can join McCormick and a long list of others in the dubiously popularized historical claims department).

JOHN F. STEWARD IN GLEN CANYON

One of the creative minds that helped build Marsh and Deering was John F. Steward. An adventurer at heart, he took the earnings from his first invention and joined the 1871 Powell river expedition into the Grand Canyon. *Courtesy of the US Geological Survey*

LEWIS STEWARD

John's older brother, Lewis, was a stalwart member of their Plano, Illinois, community. *Northern Illinois University Collection*

Plano's fertile fields drew people west in the pioneer days, and the town embraced its local agricultural inventors and entrepreneurs. The Steward family was part of this rich history. Lewis Steward, John's older brother, helped the town of Plano by inducing railroads to lay track across his land and helping get the roads properly surfaced. More importantly, he was a key part of the company that built the Marsh Harvester.

The machine was conceived during the harvest of 1857 by two brothers, Charles W. and W. W. Marsh of DeKalb, Illinois, who built a dozen or so versions before landing on a functional prototype in 1860. Lewis Steward saw great promise in the machine and invited the brothers to come down to Plano and work with his man John Hollister (John Steward's uncle) to develop it.[33]

Growing up surrounded by farm innovation, Steward was an avid reader with a voracious appetite for history that grew into a love of geology and the natural world around him. He also grew tall and strong, just under six feet with a "dark complexion and handsome bearing."[34]

Like his counterpart Legge, Steward was handsome, intelligent, good with mechanical things, and driven. In personality, however, the two were quite different. Both were straightforward in manner, but Legge had the unique ability to vocalize his often-acerbic observations without upsetting the subject of his criticism.

The equally outspoken Steward had quite the opposite talent—an innate ability to draw ire. A Deering employee described Steward as follows: "In appearance and manner he was uncouth and very difficult to get along with. He seemed to make an enemy of almost everyone with whom he came in contact."

MARSH HARVESTER

One of the innovative reapers of the 1860s was the Marsh Harvester, which was invented by Charles Wesley Marsh and his brother, William Wallace Marsh, with refinement help from the Steward family. The first patent was filed in 1858. *Floyd R. Ritzman, Northern Illinois University Collection*

CHARLES W. MARSH

Marsh lived in DeKalb, Illinois, photographed in 1894.

Floyd R. Ritzman, Northern Illinois University Collection

WILLIAM DEERING

Born in rural Maine, Deering cofounded a successful business in the east, then traveled west. He invested in a reaper, and his savvy business acumen led him to quickly grow his reaper business to be one of country's leading makers of agricultural machinery.

From William Deering Memorial Service bulletin, 1913

Steward strongly disagreed with the McCormick claim that Cyrus Sr. was the sole inventor of the reaper, and his biting prose displays his ability to draw the ire of his foes: "Popular belief that Cyrus H. McCormick invented the reaper and thus laid the foundation for Harvesting Machinery is as deeply rooted as the child's faith in Santa Claus . . . It is a half-century since Cyrus H. McCormick began advertising and claiming he was the inventor of the reaper. It has been dinned into the ears of the press, into the ears of editors, has found its way into Encyclopedias and has loaded the shelves of our libraries with false biographies."[35]

While acrid in style, Steward's book presents a mountain of carefully assembled evidence for his claim, citing law, patents, court cases, and McCormick family members, all of whom indicate quite clearly that McCormick was not the sole inventor of the reaper. While Steward was not an impartial observer—his documentation is extensive and convincing.

Legge and Steward would both contribute greatly to their times, with innovation, leadership, and integrity. Steward simply left behind a larger number of enemies.

At age twenty-one, Steward volunteered for the Union army, fighting in the Civil War with Company F, 127th Illinois Volunteer infantry.[36] During the Vicksburg Campaign, he sustained a chest wound and back injury, and he was transferred to the reserves in Michigan. He married a local woman, Sarah Louise Chandler, and returned to Plano.

After coming back, he went to work at Marsh Harvester, which was benefiting from a growing demand for farm machines as America's work force was sent to war. Like the World Wars in the twentieth century, the Civil War–induced shortages of manpower and need for higher yields accelerated the mechanization of agriculture.

Steward was the right man for the times, and his mechanical genius served him well while he worked for the local harvesting company. He invented a small device that greatly improved manufacturing of the Marsh harvester and was awarded $6,000.

He took the money and volunteered to serve on Major Powell's 1871 expedition as an assistant geologist, helping the team explore and map the Green and Colorado Rivers and the Grand Canyon. This was the second expedition by Powell, who did part of the route in 1869. Steward's war-injured back did not serve him well in this hostile environment, and six months into the expedition, he was forced to hobble home. (Powell went on to great fame. Lake Powell is named after him, and the story of the expedition was captured in books, theater, and film—including the movie *Ten Who Dared*, produced by Walt Disney in 1960.)

Steward returned to Plano, once again to work for the Marsh harvesting company, which by that time had grown, striking a deal with Elijah Gammon, who had purchased a license to build the Marsh harvesting machines in 1870.[37] William Deering, a successful businessman turned investor, put $40,000 in the company not long after. This investment would transform the fortunes of the Marsh brothers and the Steward family.

DEERING'S RAPID ASCENT

William Deering was born in 1826 in South Paris, Maine, making his business start as a boy working as a clerk in a country store.[38] He apprenticed in his family's woolen mill, and he became manager before he was eighteen years old.

Deering would seek his fortune as a farmer when he was twenty-four years old, and he found that pioneer life wasn't his cup of tea.[39] He returned to the family mill in 1856. During the Civil War, Deering secured a contract to supply uniforms to Union soldiers. He earned enough in that transaction to have some money to invest. When the war ended, he and a partner, Seth M. Milliken, founded the Deering Milliken Company, a woolen fabrics distributor in Portland, Maine. They moved the company headquarters to New York City in 1868.

Deering traveled the country looking for investments and encountered Elijah Gammon, who was from Maine and was an acquaintance of Deering. Gammon had a license to build harvesters designed by the Marsh brothers. Deering invested $40,000 in Gammon's business. Two years later, the company showed $80,000 of profit.[40]

In 1873, Gammon "fell sick and begged Deering to come to Illinois and manage the business."[41] This series of events put the retired Deering into a completely foreign business—one so new to him that he told Herbert Newton Casson, author of *Romance of the Reaper*, that he did not even know what the machines looked like.

He also happened into the business during the Jay Cooke Panic of 1873—a perilous time when a large bank heavily vested in railways (Jay Cooke & Company) went bankrupt. The collapse of the bank led to a nationwide panic. As people across the country withdrew their funds, more than one hundred banks and eighty-nine railroads went bankrupt, and the country's economy plunged into depression.

To compound matters, competition between the agricultural companies was at a fevered peak. This was the time of the "Reaper Wars," when marketing consisted of field demonstrations that were host to marketing gimmicks, sabotage, rigged contests, subterfuge, and fisticuffs.

The legendary Reaper King, William Whiteley, the flamboyant figurehead of the Champion brand of reapers, hitched

GAMMON & DEERING

Deering invested $40,000 with his acquaintance, Elijah Gammon, who had a license to build Marsh harvesters. *Wisconsin Historical Society / 4280*

PANIC OF 1873

When one of the largest banks in the country collapsed on September 20, 1873, the American people panicked, and the stock exchange collapsed. The front page of this newspaper shows the stock exchange doors being closed at the fateful day's end. *Library of Congress LC-USZ62-37423*

himself to his machine at Jamestown, Ohio, in 1867. He pulled a "full swath with no horses, just his broad back strapped into the harness."[42]

A salesman for Deering, J. W. Wisehart, recalled the field trials as all show, and no trial: "During the early years of my harvester experience, the field trial was an important part of the sales argument to the farmer. We never entered a trial unless we had it fixed and won beforehand. That was the idea of them all."

Deering's gift was creating profitable businesses, and he grew his company rapidly. He also was quick to invest in new technology. He was blessed with the Marsh harvester—which was a good machine—and with Steward, who was a genius in his own right.

The company did make a mistake or two—notably investing in adding a wire-binder to the Marsh machine for the 1868 harvest season. The binders failed to work properly, and they had to be returned, rebuilt, and resold the following season. Deering would go on to find another farmer-inventor that proved key to his company's future.

THE APPLEBY BINDER

A southern Wisconsin farmer-inventor, John F. Appleby, designed a device that bound grain by wrapping it in twine and tying it off with a knot, a significant improvement over the wire binders of the time. He was eighteen years old when he built his first working example. He served in the 23rd Wisconsin Volunteer Infantry Regiment in the Civil War, and when he returned to Wisconsin at age twenty-four, he perfected his binder. He had it patented four years later, in 1878.

Early that same year, Steward and Deering were invited to see the Appleby binder at work at the factory of Parker & Stone in Beloit, Wisconsin—a company that held the license to build the machines.[43] A deal was struck, and two binders built by Parker & Stone were shipped to Texas to be tested by Steward for the fall harvest. Parker & Stone felt it lacked the backing to promote the machines, and, on November 29, 1878, Deering purchased the license for the Appleby binder. With help from Appleby, improvements were made in the ensuing two years. Casson states that the final improvements were made by Steward.

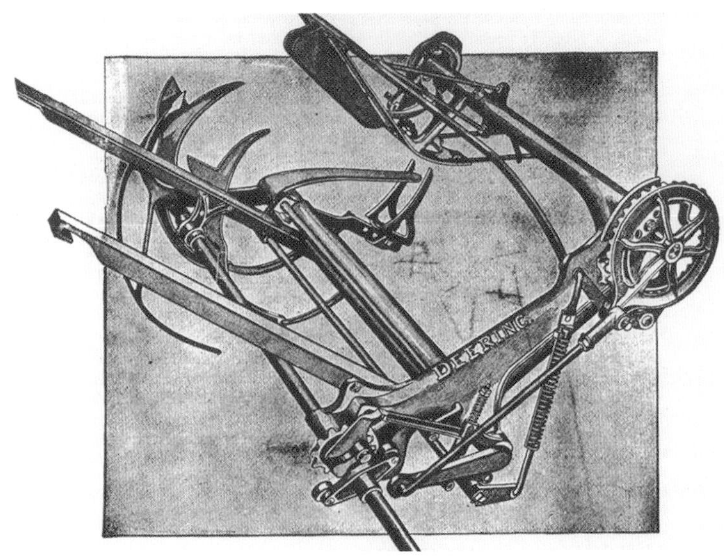

APPLEBY TWINE BINDER

Once the reaper had been refined to cut grain efficiently, the next big advancement was binding the grain into bundles. Deering and Steward discovered a young man, John Appleby, who invented an excellent twine binder. Deering licensed the patent, bet his fortune, and reaped the reward. *Robert L. Ardrey, Northern Illinois University Collection*

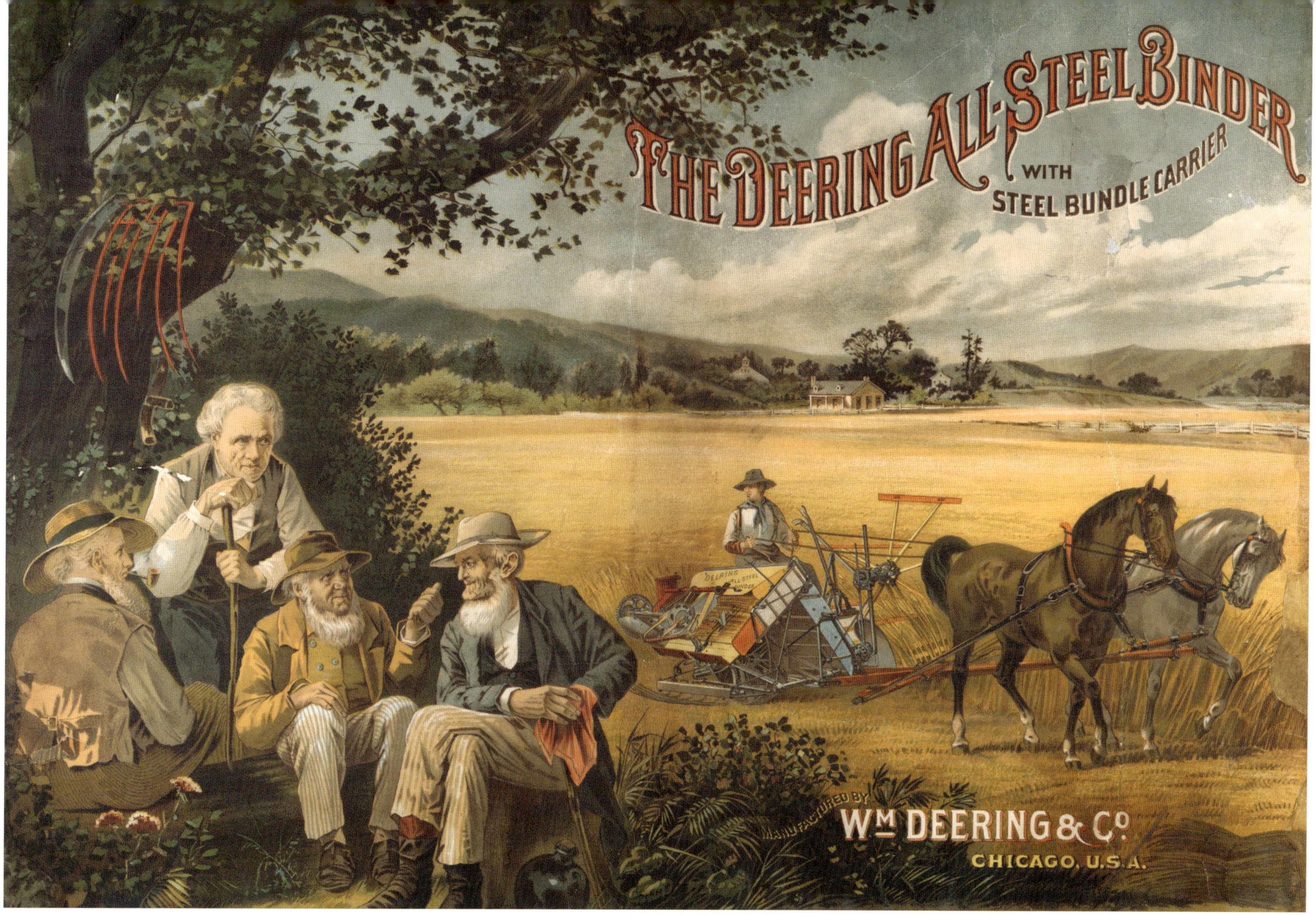

MATERIALS AND MARGIN

After the binder was perfected, the next advancements were more affordable twine and lighter, stronger machines, as seen in this 1883 catalog cover. Deering pioneered on both fronts and it was also one of the first to purchase lands with good ore prospects. By the end of the nineteenth century, it was one of the most profitable agricultural equipment makers.

Wisconsin Historical Society / 4386

In 1880, Deering built three thousand of the machines for the harvest season. The Casson book claims Deering bet his life's fortune on the venture. Gammon sold out to Deering in 1880, allegedly because he was reluctant to invest and build the three thousand twine binding machines. Deering's twine binder was a tremendous success, netting him a reputed $400,000 in profit and laying a cornerstone in the Deering ascension. His other stroke of genius was to reduce the cost of twine, which was part of the reason Deering led the reaper industry with an 18 percent profit margin.

With gross revenues quickly ballooning to near the level of the McCormick company, Deering established himself as one of the leaders in the reaper business.

THE McCORMICK EMPIRE

As the Deering company was growing, the McCormick company was solidifying its position as the market leader. By the time Deering was forcing McCormick's hand with its fancy new twine binder, the McCormick children were coming of age.

At age forty-nine, Cyrus Sr. had married his twenty-three-year-old secretary, Nancy "Nettie" Fowler McCormick, and they had their first child, Cyrus Jr., the following year, followed by Mary, Anita, Harold, and Stanley (several of their children lived only a few years).

His late marriage had a significant effect on the McCormick business evolution. His sons were young when it was time for them to take over, and his much-younger bride, Nettie, would use her wisdom and strongly opinionated hand to become one of most influential McCormick family members in company history.

After his marriage, the state of his business was sound despite a cold short season in 1856 and 1857 as well as an economic recession in 1858 and 1859. Farmers were in dire straits, with falling grain prices forcing an increase in acreage

NANCY "NETTIE" FOWLER McCORMICK (1835–1923)

Married to Cyrus Sr. in 1858, Nettie would become one of the most influential people in the McCormick company, and perhaps more so after the merger. She was twenty-seven years old in this image made in 1862. *Matthew Brady, Wisconsin Historical Society / 60645*

WILLIAM SANDERSON McCORMICK (1815–1865)

The older brother of Cyrus Sr., William worked with him to build the McCormick company. An avid outdoorsman, William preferred to be in the fields of their Virginia home or hunting in Wisconsin or Minnesota. His gifts for fiscal management meant he spent much time doing the "inside work" he disliked. *Wisconsin Historical Society / 60639*

LEANDER JAMES McCORMICK (1819–1900)

The younger brother of Cyrus Sr., Leander also worked to build the McCormick company. He and Cyrus Sr. would struggle at times over the years, disagreeing over finances and their father's legacy and role in the invention of the reaper. *Century Publishing and Engraving Company, Wisconsin Historical Society / 8389*

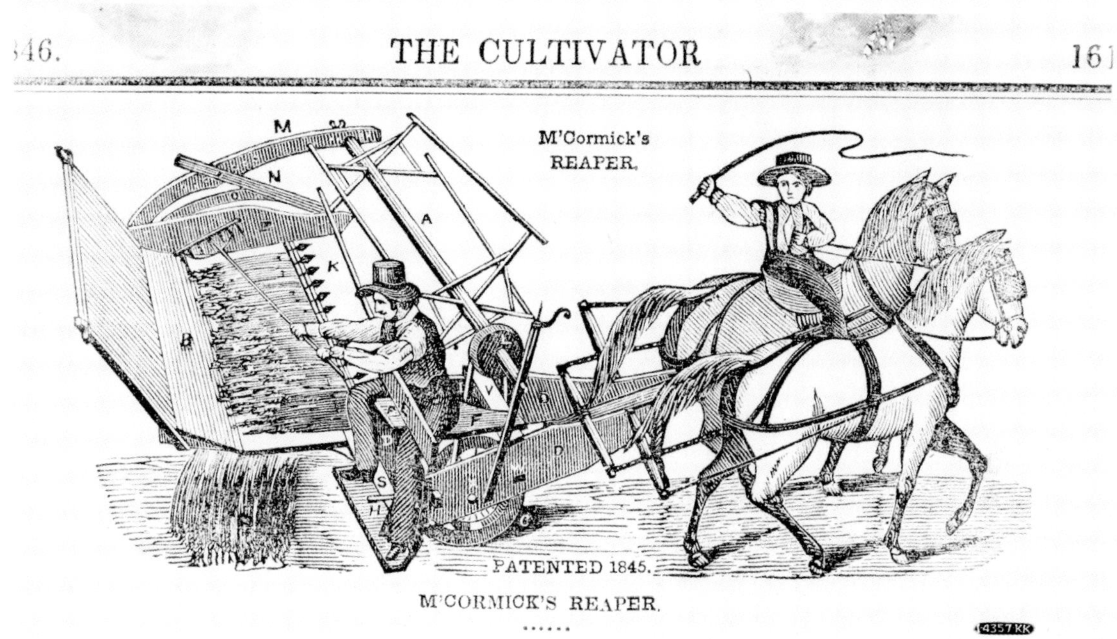

CULTIVATOR-REAPER DIAGRAM (1867)

This rendition of the reaper requires a man in the back to rake the stalks. Ironically, this diagram was probably stolen from Walter A. Wood, a competitive reaper company, as the envelope was typewritten with, "Copied from Walter A. Wood." *Wisconsin Historical Society / 3340*

harvested to make ends meet. The ideal way to increase output was with farm mechanization, and the McCormick company sold every reaper they could build from 1856 to 1860. These were sold on credit at the time, so moving the machines was not the hard part. Collecting the money could be very difficult, and some companies failed during this time due to inadequate capital, poor collections, or both. McCormick's policies of pushing his salespeople to carefully vet the farmers they sold to, and then to avoid taking anyone to court for payment, paid off well in this era. McCormick's brand reputation was better than those that sued a lot of farmers, and the customers they selected of better quality.[44]

As the McCormick company grew more profitable, the family invested heavily. Cash was extremely volatile and investment a wiser choice. Chicago real estate was cheap, with great rates of return due to the city's growth, and the family invested heavily in both farm and city property. The family in fact became the "largest landlord in Chicago."[45] Cyrus Sr. also purchased interests in banks, lumber and coal rights, and more.

Cyrus Sr.'s brothers, Leander and William, were active members of the company throughout the 1850s. William managed much of the company's finances. He had a flair for risk and a nose for good deals, and the investments he managed did well.[46] As the decade drew to a close, the two men began to forcefully express the belief that their contributions outweighed their compensation. In 1857, William told Cyrus Sr. that he expected more than just a salary, and Leander threatened to quit unless he received a raise. Cyrus Sr. relented on both accounts, and he would make another major concession in 1864.

In July 1860, Cyrus Sr. purchased two newspapers, the *Chicago Herald* and the *Chicago Daily Times*, and combined them to create the *Daily Chicago Times*. This was one investment that did not pay off. Circulation never surpassed three thousand copies, and the paper was not fiscally successful. In April 1861, he announced he would turn over operations of the paper to his brother-in-law, Eldridge M. Fowler. By June 1861, Cyrus Sr. sold the paper to Wilbur F. Storey. Under Storey's leadership, the paper's circulation grew to thirty-five thousand. After Storey died, the paper was merged with the *Chicago Herald*.[47]

THE CHALLENGES OF THE CIVIL WAR

In 1861, the American Civil War created huge business challenges. McCormick couldn't meet demand that year and in 1862, estimating that more than 2,500 sales went unfilled due to shortages (a typical production run at this time was about five thousand machines). In 1863, low crop prices and difficulties delivering machines left the company with a surplus of two thousand machines. Prices fluctuated wildly throughout the war, and, by the end, farmers became understandably wary of new equipment purchases.[48]

As with later wars, the McCormick company struggled with raw material supplies and costs. Lumber prices doubled, and the cost of coal and pig iron nearly tripled. Hiring enough skilled labor during this time was challenging, with strikes also causing production delays. Transportation costs increased on the rails and the water, and at times routes were shut down due to battles or closed borders. Currency could also be an issue—the northern banks would sometimes refuse to accept the Confederate dollar.

Competition for sales were fierce, making large price increases suicidal. McCormick only increased prices by roughly 15 percent, despite the huge cost increases. In 1862, the Eastern Seaboard harvesting companies, including McCormick, met and agreed to increase prices 10 percent for reapers and mowers. Meetings like this continued into 1864, with much discussion of industry-wide price increases and terms. The business practice was collusion, and it would be made illegal by the Sherman Act of 1890.

Despite all these challenges, the Civil War increased farmer's reliance on and use of farm machinery, particularly reapers that replaced the handwork of cutting grain with a cradle scythe.

REAPERS IN THE CIVIL WAR

When the Civil War broke out in 1861, the McCormicks were keenly aware each machine freed up two or three farmhands to fight. They sold 250,000 reapers and mowers during the Civil War. The company presented their machine as battlefield contributors with a series of posters like the one you see here. *Cosack & Co., Library of Congress LC-DIG-pga-00540*

1. General Grant and Staff.

OF SHILOH · APRIL 6th 1862 ·

-MICK" MACHINES COME VICTORIOUSLY OUT OF EVERY CONTEST, AND WITHOUT A SCRATCH.

· PRESENTED WITH COMPLIMENTS OF · McCORMICK HARVESTING MACHINE COMPANY ·

COPIED BY SPECIAL PERMISSION From The · PANORAMA PAINTING · ON EXHIBITION IN CHICAGO.

POST-WAR PATENTS AND MORE

In September 1864, William succumbed to dysentery of a typhoid nature. Before he died, he made peace with Cyrus Sr. and urged his brothers to "forbear one another in love!"[49]

Cyrus Sr. was devastated by the loss. William had worked with him since the beginning. In addition to his skill with managing money and business, he was the peacemaker in the family. Cyrus Sr. and Leander would strike a new deal in 1866. The two would argue over the value of real estate holdings William willed to Leander—mainly large tracts of farmland—and the relationship would eventually cool.[50]

As the nineteenth century progressed, farm technology evolved as did the methods of doing business. Where early entrants such as Hussey, Manny, and McCormick battled mainly with their own innovation, the game changed as time wore on. The nascent patent law of the era was evolving rapidly and was incredibly complicated. Most of the innovation in the middle and latter part of the century were improvements upon existing machines, with patents typically awarded to "farmers or to small-town mechanics."[51]

CYRUS HALL McCORMICK (1809–1884)

Cyrus Sr. was politically active, particularly later in his life. He helped found the YMCA, donated extensively to several Lutheran and Presbyterian colleges, and supported efforts to establish churches and Sunday schools in the South. He ran unsuccessfully for Congress in Illinois's First District in 1864. *Library of Congress LC-USZ62-27710*

These patents almost invariably ended up in the hands of one of the large manufacturers—the Appleby twine binder's ownership by Deering being a prime example. It was a critical development, and Deering and Steward were wise enough to snap it up. In fact, Steward alleges that the McCormick company took the machine apart and copied the design.[52]

William T. Hutchinson, author of the two-volume biography of Cyrus Sr., acknowledges this was common practice when innovation occurred. "McCormick's agent, acting incognito, would purchase and ship to Chicago one of the machines which had been so successful in the season just closed. The inventors at the factory then studied it carefully for the purpose of discovering a way whereby the implements in their charge might maintain a similar perfection of operation, without making their employer liable for a suit for infringement of patent-rights."[53]

If the innovation was a game-changer and no way could be found to recreate it without a violation, then the company would be forced to pay to license it from the patent rights holder. Royalties for patents could often be as lucrative as selling machines. Scouting for new inventions and managing patent rights was, at the time, a critical part of maintaining a business in this very competitive field.

Lawsuits were also a useful tool. Beyond the obvious reason of stopping patent infringement, the courts could be wielded to deter or delay competitors from using innovations. Lawsuits were also a good way to draw attention to your product in newspaper headlines and handbills, pamphlets, and posters. "It was a quiet year when the McCormicks, the Deerings, or Walter A. Wood were not parties to three or four lawsuits."[54]

In this era, Cyrus Sr. devoted much of his time to innovation of business management rather than machinery. Given the complexity of patent right acquisitions, royalty payments for and from patents, and legal fees, projecting business profit and loss was difficult to do and vital to company success. As the most successful farm equipment business of the era, Cyrus Sr. and the McCormick company were superb at business management.

1899 illustration, Puck *magazine*

In the mid-1860s, Cyrus Sr. dabbled with investment in more risky ventures. "He relished a new financial adventure and enjoyed it as long as it was not too expensive. Participation in hazardous schemes afforded him a release from the humdrum affairs of every day."[55]

Mines and railroads were of particular interest, and Cyrus Sr. invested in the Union Pacific (via a dubious investment firm, Crédit Mobilier), the Southern Railroad Association, the Richmond & Allegheny, the Schoolcraft mine in Michigan, the ill-fated Dorn Mine near McCormick, South Carolina, and others. Hutchinson concluded that Cyrus Sr. suffered a modest loss on these high-risk ventures, writing: "As a speculator, McCormick was not a success."[56]

THE BUSINESS OF THE COURT

Patents and patent litigation was immensely complex in the late 1800s. Savvy company leaders at the time used both to generate income as well as garner free publicity from newspaper coverage of the court battles.

Wisconsin Historical Society / 4415

RISING FROM THE ASHES

When the Great Chicago Fire of 1871 destroyed the McCormick factory, Cyrus Sr. reputedly expressed great doubt about rebuilding. The loss was valued at $2 million[57]—one of the largest in the Chicago fire and $48 million in 2023 dollars—and the fire destroyed the company's patterns and machinery. Legend has it that Cyrus Sr.'s concerns were overruled by Nettie, who insisted they rebuild. They did so in fine fashion.

Construction began on the new McCormick Reaper Works in 1872. The plant was built on 155 acres on the north branch of the river, with 1,320 feet of dock footage. The buildings alone covered 21 acres of ground and were equipped with all the latest machinery.

The McCormicks built not only the works, but also homes, office buildings, and other buildings in Chicago in the ensuing years.

The new factory enabled McCormick to cement its place as a leader in the agricultural implement industry. Farm equipment was built in the factory and on that ground—with more than fifty-three buildings on it—until McCormick Works closed in 1959.

GREAT CHICAGO FIRE

On October 8, 1871, fire broke out in the southwest of Chicago. Roughly two-thirds of the city's buildings at that time were wood, as were the sidewalks. The fire spread quickly and engulfed 2,000 acres. The blaze destroyed 17,500 structures in Chicago, including the original McCormick factory. *Currier & Ives, Library of Congress LC-DIG-pga-06548*

McCORMICK WORKS 1887

The McCormicks rebuilt their factory on 155 acres on the north bank of the Chicago river, featuring 1,320 feet of water frontage. The works itself covered 21 acres and remained in place until 1962. *Wisconsin Historical Society / 39554*

McCORMICK REAPER AND BINDER

By the 1880s, reapers were able to bind and cut grain, with no need to have anyone walk with it and tend to the horses or the grain. This model was built in 1883 and added a binder to the operation, but it used a frame that was largely made of wood. *Wisconsin Historical Society / 22888*

McCORMICK FAMILY TREE

Robert Sure McCormick II (1780–1846) m. **Mary Ann Hall** (1780–1853), 1808

- **Robert Hall** (1810–1826) d. at age sixteen
- **Susan Jane** (1813–1826) d. at age twelve
- **William Sanderson** (1815–1865) m. **Mary Ann Grigsby** (1828–1878), 1848
- **Mary Caroline** (1817–1888) m. **James Shields** (1812–1862), 1847
- **Leander James** (1819–1900) m. **Henrietta Hamilton** (1822–1899), 1845
- **John P.** (1820–1849) d. at age twenty-eight
- **Amanda Joanna** (1822–1891) m. **Hugh Adams** (1820–1880), 1845
- **Cyrus Hall Sr.** (1809–1884) (1) m. **Nancy Maria (Nettie) Fowler** (1835–1923), 1858
    - **Cyrus Hall Jr.** (1859–1936) (2) m. **Harriet B. Hammond**, 1889
        - **Cyrus Hall III** (1890–1970) m. **Dorothy Linn**, 1915
    - ------- m. **Alice Marie Hoit** (1881–1950), 1927
        - **Elizabeth** (1892–1905) d. at age twelve
        - **Gordon** (1894–1967) (3)
    - ------- m. **Florence Davey** (1888–1979), 1931
    - **Mary Virginia** (1861–1941)
    - **Robert** (1863–1865) d. at age fourteen months
    - **Anita Eugenie** (1866–1954) m. **Emmons Blaine**, 1889
        - **Emmons Jr.** (1890–1918). m. **Eleanor Gooding** (1893–1972), 1917
            - **Anne (Nancy)** (1918–1977) m. **Gilbert Harrison** (1915–2008), 1951
            - **Emmons Blaine** (1918–1918)
    - **Alice** (1870–1871) d. at age eight months
    - **Harold Fowler** (1872–1941) m. **Edith Rockefeller** (1872–1932), 1895, div. 1921
        - **John Rockefeller** (1897–1901) d. at age three
        - **Harold Fowler Jr.** (1898–1973) (4) m. **Anna Stillman**, 1931
        - **Muriel** (1902–1959) m. **Elisha Dyer Hubbard**, 1931
        - **Editha** (1903–1904) d. at age eight months
        - **Mathilde** (1905–1947) m. **Wilhelm Max Oser** (1877–1942), 1923
            - **Anita**
            - **Peter Max**
    - ------- m. **Ganna Walska** (1887–1984), 1922, div. 1931
    - ------- m. **Adah Wilson**, 1938
    - **Stanley Robert** (1874–1947) m. **Katherine Dexter** (1875–1967), 1904

1. Commonly signed his name as **C. H. McCormick**.
2. Originally named **Cyrus Rice**, in his youth changed to **Cyrus Hall**, and for many years thereafter signed his name as **Cyrus H. McCormick, Jr.**
3. Never married.
4. Familiarly known as **Fowler**.

McCORMICK'S NEXT GENERATION

By the time Cyrus Sr. passed away in 1884, he had built an enormously successful company, with a broad line of agricultural equipment sold and marketed worldwide. His most vocal critic, John F. Steward, wrote: "His exceeding conservatism made him money, as he spent comparatively little in experiments. He adopted what others had proved to be valuable."[58]

Cyrus Sr. built his business with a savvy understanding of manufacturing processes and product promotion, as well as a dealmaker's appreciation of sales techniques and the critical nature of extending credit to farmers (and then following through with effective collections techniques). This blend of talents was unique and outstanding and earned his heirs the resources necessary to cement his name into legend.

Cyrus Sr.'s widow, Nettie, had become one of the most trusted personas in the business. She was consulted on important business deals for decades after Cyrus passed. Her son, Cyrus McCormick Jr., took over the helm at McCormick after his father passed. He was twenty-five years old at the time, his only experience in business having been "five years . . . as his father's confidential secretary and representative."[59] While Cyrus III described his father, Cyrus Jr., as "calmer" and "more adaptable" than Cyrus Sr., he was also young and inexperienced.

Leadership of the company at this time was complicated by the fact that family disagreements over compensation and recognition were boiling over. Cyrus Sr.'s brother, Leander and his son, Robert, became increasingly aggressive and litigious in their efforts. They wanted more pay—Leander for a time received only a modest salary—and they felt the family patriarch, Robert, should get more recognition for his work developing the original mechanical reaper. They also disagreed over patents.[60] After a settlement in 1879—and Leander leaving or being ousted, depending on who you believe, Leander and Cyrus Sr.'s relationship improved a bit. After Cyrus Sr. passed away, however, the family split was magnified.

At twice Cyrus Jr.'s age and a formidable force, Nettie undoubtedly provided a strong guiding hand as the young man grappled with the challenges of leading a very large corporation. Cyrus III wrote, "The younger McCormick also had his best ally the unfaltering support of his mother. Had she lived in other times, Nettie F. McCormick would have headed the business herself."[61]

CYRUS HALL McCORMICK JR. (1859–1936)
When Cyrus Jr. was twenty-five years old, his father passed away and the young man stepped in to take the helm of the McCormick company. This portrait of him was made in 1880. Max Platz, *Wisconsin Historical Society / 8371*

NETTIE AND CYRUS SR.
When Cyrus Sr. passed away in 1884, his widow, Nettie Fowler McCormick, was forty-nine years old. *Brands Studios, Wisconsin Historical Society / 8775*

THE HAYMARKET SQUARE RIOT

One of the most important events in labor history was a battle that took place between management and workers at the McCormick Reaper Works in Chicago in 1886.

Trouble started brewing at the plant as far back as 1877, when protestors and policemen scuffled. Things ramped up again shortly after Cyrus McCormick Jr. stepped in to replace his father as head of the McCormick company. At McCormick Reaper Works, an April 1885 proposal for a 15-percent wage decrease led to a bitter strike that left the young man determined to break the union.[62] "The right to hire any man, white or black, union or nonunion, Protestant or Catholic, is something I will not surrender," Cyrus Jr. stated.[63]

Cyrus Jr. had no intention of letting the union into his operation and introduced new machinery that eliminated skilled workers. In February 1886, he locked the union out of McCormick Works and replaced them with nonunion workers. The conflict grew increasingly heated, and three hundred armed Pinkertons—private detectives—were hired to protect the nonunion employees from angry union demonstrators.

On May 1, a nationwide strike was organized in which three hundred thousand workers in thirteen thousand establishments stopped work to push for an eight-hour workday. The day passed uneventfully.

On May 3, 1886, well-known Chicago-area activist and publisher August Spies spoke at a gathering of several thousand Lumber Shovers Union members—a meeting that took place a few blocks from the McCormick plant. As Spies spoke, the workday ended at the plant and the crowd began heckling the nonunion replacements who were leaving for the day. A melee erupted, and seventy-five policemen and a paddy wagon came in to break it up. The dissidents threw stones, and the cops started firing into the crowd. At least two demonstrators were killed.

The next day, Spies distributed a small handful of circulars, touting the fact that six of their brothers had been killed. A demonstration was called for in Haymarket Square at 7:30 p.m. on the evening of May 4.

When Spies arrived at 8:15 p.m., only a few thousand people were assembled. He spoke to the crowd, calling for them to focus on the eight-hour workday, and addressing the

THE CATALYST

On May 3, 1886, the night before the Haymarket Square Riot, policemen and protesters clashed near the McCormick Reaper Works. Several demonstrators were killed. *Charles Upham & C. Bunnell, Northern Illinois University Collection*

THE ACTIVIST

This illustration depicts publisher August Spies speaking at the McCormick Works on the night of May 4. He and several others who spoke at the event would be tried for the murder of policeman Matthias J. Degan. *J. Schutte & Company 1889, Northern Illinois University Collection*

HAYMARKET SQUARE RIOT
On the night of May 4, 1886, a bomb was thrown at a group of policemen at a labor protest, and the police began firing into the crowd. Sixty-seven people were hurt in the fracas, and seven of them died. *Library of Congress LC-USZ62-75192*

fact that Cyrus Jr. had personally blamed Spies for the being the instigator of the violence. He brought up Albert Parsons, who spoke until 10 p.m., and Samuel Fielden spoke next.

While Fielden spoke, a large group of police officers came into the square and ordered the crowd to disperse. Shortly thereafter, someone threw a bomb into the midst of the policemen, who drew revolvers and began to fire into the crowd. Seven policeman and an unknown number of civilians were killed in the incident. Sixty policemen were injured. Officer Mathias J. Degan had his leg sliced open by a fragment of the bomb and bled to death in the street. The identity of the bomb-thrower was never established.

Two of the men who spoke at the meeting—Spies and Parsons—were sentenced to death for the murder of Degan. Along with two others, they were hanged on November 11, 1887. The incident lives on as a dramatic example of labor's struggle with powerful industry leaders and is viewed as a terrible miscarriage of justice.

ANARCHISTS EXECUTED
This illustration in *Frank Leslie's Illustrated Newspaper* showed the four convicted demonstrators from the Haymarket Square Riot being led to be hung, including publisher and activist August Spies at the head of the line. The other three men were Adolph Fischer, George Engel, and Albert R. Parsons. *Library of Congress LC-USZ62-109367*

The Haymarket Square Riot impacted how the McCormicks, and later International Harvester, dealt with labor for decades. While there were struggles over the years, company leaders were generally cautious and fair with labor negotiations during the first part of the twentieth century. Ironically, IH leader Archie McCardell would deal harshly with labor in 1979, with calamitous results.

WORLD EXPOSITIONS

The early conflicts between reaper manufacturers were carried in more genteel fashion at the nineteenth century's World Expositions. Also referred to as World's Fairs, the events were large technology exhibits that stayed open for up to six months.

The Grand Exhibition held in London's Hyde Park in 1851 showcased more than fourteen thousand exhibitors of new technology from around the world. Both McCormick and Hussey reapers were displayed. The *London Times* mocked McCormick's reaper as "a cross between an Astley Chariot, a wheelbarrow, and a flying machine." The American reapers would have the last laugh, winning more prizes than the British variants, prompting the *Times* to write: "This machine is worth to the farmers of England the whole cost of the exhibition."[64]

McCormick would continue to display his machines at World Expositions, winning awards in most of the major shows worldwide. These displays helped grow their international business as well as identify promising new markets.

1851 CRYSTAL PALACE

The Great Exhibition was held in London's Hyde Park from May 1 to October 1 in 1851. A massive glass house 1,848 feet long was built for the event. Charles Darwin, Charles Dickens, and Samuel Colt were among the six million people who attended. *Dickinson Brothers, via Wikimedia Commons*

NETTIE AND CYRUS McCORMICK JR. AT 1900 EXPOSITION UNIVERSELLE

The response to their machines at the 1900 Exposition Universelle in Paris elicited new interest in global expansion. Unfortunately, they were unable to raise enough additional capital to do this. This issue is cited as key to the McCormick interest in a merger. *Wisconsin Historical Society / 8383*

McCORMICK AWARDS

The McCormick company maintained its presence at Grand Expositions throughout its history. *Wisconsin Historical Society / 54695*

THE COWBOY AND THE PRINCESS

By 1895, our man Alexander Legge was continuing to distinguish himself in collections for McCormick in Nebraska and was in fact the lead man in the territory. The "tall cowpuncher" made his mark with tough but fair negotiations.

A favorite tale is of the farmer who put up his bull for collateral when he purchased his equipment. When Legge requested the bull be returned to offset missed payments, the following conversation ensued.

"Where is the bull?"

"Out there in that little pasture."

"I don't see any bull there."

"Of course you don't," the farmer said. "He's under the ground not on top of it—dead as a mackerel."

To which Legge asked for a couple of shovels. "His bones will bring a little something at the rendering plant."

The farmer was silent for a few moments, then replied, "Hell! If you're that kind of collector, I might as well pay up!"

Legge would also go to bat for the farmers, getting notes paid by visiting their local banker and enticing them to extend credit to his customer, or helping a farmer employ a bookkeeper to straighten matters so he could pay his bill.

A marriage that would make headlines would also impact our cowboy collector. Harold F. McCormick—the son of Cyrus Sr.—would marry Edith Rockefeller, the daughter of John D. Rockefeller—founder of Standard Oil.

The connection of the young couple was made by Harold McCormick, who had met John Rockefeller Jr. while they both attended the Browning School, a highly selective boys prep school in New York.[65] John, his wife Cettie, and three of his sisters visited Harold in 1893 and stayed at Nettie's house. Harold met Edith there. The two had both grown up in wealthy households in which strict limitations were placed on the children, with strong value placed in charitable work and a relatively modest lifestyle.

"The Rockefellers deplored the vogue among rich Americans of marrying off their daughters to title Europeans and welcomed the McCormicks as an upright, God-fearing industrial family," Ron Chernow wrote in *Titan*, going on to describe Harold as an "athletic man with luminous blue eyes and a dreamy gaze who wore jeweled cufflinks and embroidered vests . . . he stood out for his free and open manner." Rockefeller Sr. took a shine to Harold, and he was the only son-in-law permitted to smoke in John's presence.

Rockefeller Sr. also abhorred drinking, and repeatedly appealed to Harold to refrain, and attempted to enlist Edith in his case. This was to no avail.

Harold was twenty-three years old when he and Edith were married in 1895, and he took an assignment to work for the family business in Council Bluffs, Iowa.

Alexander Legge was assigned to teach Harold the ropes. He also was assigned to outfit the couple's new home. While the couple was not terribly impressed with his decorating skills, they were taken in by his frank, no-nonsense approach, impeccable decision-making, and personal integrity.

"He was refreshingly simple and honest, and engagingly human, as great of heart as of mind."[66] Harold said of Legge, "He was a gentleman in the most vital definition of the term. His rough exterior covered a rich endowment of fine feeling and natural discrimination in human relationships . . . his reasonings and his decisions were based on practical experience and close personal observation."[67]

Nettie Fowler McCormick later wrote, "Whatever Mr. Legge says is the law."[68]

In 1899, Harold had moved from Council Bluffs to Chicago to take on the role as the first vice president of the McCormick company, sitting at the right hand of his brother, Cyrus Jr.

COUNCIL BLUFFS, IOWA

The town of Council Bluffs, photographed in 1916.

Library of Congress LC-USZ62-127715

HAROLD McCORMICK AND EDITH ROCKEFELLER

Harold and Edith were twenty-three years old when they got married in 1895. After a long honeymoon abroad, these children of great wealth would start their life together in Council Bluffs, Iowa. *Wisconsin Historical Society / 8374*

ALEXANDER LEGGE

Legge was tasked with helping young Harold take on a leadership role, first in the local branch and then at the national level of the McCormick company. He is about twenty-two years old in this image.
From Alexander Legge 1866–1933

Harold sent Legge a note, asking him to come to Chicago to take charge of the company's collection business. Legge quickly distinguished himself with the acuity, tact, and intelligence that served him well in Nebraska.

Legge would be one of the McCormick family's most trusted lieutenants—and go on to play a variety of key roles for several decades.

When Legge went to the McCormick offices in Chicago, the winds of change were already brewing, with events percolating that would propel a Nebraska farm boy and ranch hand with minimal education to leader of one of the world's largest companies.

IOWA FARM COUNTRY

Farm scenes in rural Iowa.
Library of Congress

THE MERGER

Discussions of merging harvester companies began in 1885 and became serious late in 1890, when Colonel A. L. Conger, president of Whitman & Barnes Manufacturing Company, cooked up a scheme to bring twenty of the harvesting companies under one roof.

His idea was met with enthusiasm, and on November 19, 1890, the new American Harvester Co. was incorporated in Illinois with McCormick, Deering, and eighteen other founding companies. Contracts were sent out with retail prices, and some employees were let go. When bankers became involved, the deal soured.

Cyrus McCormick III wrote that Deering Sr. and McCormick Sr. agreed to back out after a late-night meeting in a New York hotel lobby.[69] Harold McCormick stated that the key moment was when Deering Sr. visited Nettie and indicated he wanted out of the deal.[70] The McCormicks agree that the deal fell through because of a lack of capital. Lawyers advised that the entire agreement was "void as being in restraint of trade."[71] Whatever the cause, the American Harvester Company lasted all of a few weeks.

CHARLES SCHWAB WITH ELBERT HENRY GARY

Judge E. H. Gary (right) founded US Steel. He would play a significant role in the harvester merger and serve on the board for a time during the early years of the company. *Library of Congress LC-DIG-ggbain-08608*

In 1897, the McCormicks seriously discussed purchasing Deering outright—but Deering wanted too much money, so that was scrapped. Another failed attempt to merge was made in 1901.

The steel industry—notably Wisconsin Steel—plays a role in the International Harvester Company's modern history, and it would also factor into the company's foundation.

In February 1902, E. H. Gary of US Steel called Cyrus Jr. to discuss the disturbing trend he had noticed in the harvester industry.[72] Deering, in particular, was buying up ore lands and building a rolling mill. Likely concerned with this intrusion upon the steel industry, Gary suggested that rather than buy up the supply line soup to nuts, McCormick should consider improving margins by establishing a harvester trust.

He outlined a plan to McCormick that would allow the newly combined companies to multiply their value by issuing both common and preferred stock. The scheme allowed trusts to, in general, increase the value of their holdings by about 35 percent. Gary explained that someone like J. P. Morgan could facilitate such a thing.

Cyrus Jr. was intrigued and discussed the idea with his brothers and Nettie. They liked the idea but were wary of dealing with Morgan. They certainly knew him by reputation—and the Rockefeller family had dealt with Morgan extensively. Rockefeller Sr. had a mild to openly adversarial relationship with J. P. and his firm.[73] In April and May 1902, discussions were held between the Deering and McCormick families in Florida and Washington, DC.

Plano's owner, J. J. Glessner, was aging and ready to leave the business. He ran into Harold at a meeting of Chicago's Orchestral Association, and told him wearily, "If only the Deerings and the McCormicks could get together, the rest . . . would be really easy."[74]

In June, the McCormicks held a family conference at the Waldorf Astoria in New York, and, with advice from financiers and lawyers, discussed six different plans to merge the harvester companies. There was great concern about raising the ire of the government for anti-trust violations, and several plans were rejected because they clearly would cause legal issues.

The McCormicks were considering having Francis Lynde Stetson, a lawyer they had worked with in the past who was with them that June in New York, handle the deal. On June 13, Cyrus Sr.'s son, Stanley McCormick, and relative Eldridge M. Fowler called on John D. Rockefeller Jr. to ask his opinion of Stetson.

On June 14, Rockefeller Jr. called George Perkins, the right-hand man to J. P. Morgan, and asked his opinion of Stetson. Perkins replied that Stetson was a lawyer of finest caliber, but not well-suited to manage a merger. He suggested that the McCormicks consider working with himself and J. P. Morgan.

That afternoon, Cyrus Jr. and Stanley "hurried from the Standard Oil Building on Broadway to the Morgan offices on Wall Street"[75] to meet with Perkins. He briefly outlined his thoughts and suggested a follow-up meeting the next Monday afternoon. They reconvened at the Waldorf-Astoria and Perkins outlined his plan.

He explained that the negotiation would be done by the Morgan firm, and that these types of deals unfolded organically and not by one fixed path. He thought the price for Morgan to put together the merger would be roughly 1 percent of the new company valuation. After speaking with Morgan and determining they would need to form and potentially provide fiscal backing for the new company, he later increased his estimate to $2 to $5 million. He also said if additional capital was required—as was often the case with large mergers and trusts at this time—Morgan would put up the capital in exchange for stock.

As Stanley, Harold, and Cyrus Jr. warmed up to Perkins's idea, they shared the fact that they had a harvester company ready to purchase. The McCormicks had employed P. D. Middlekauf as a secret agent to negotiate the purchase of the Milwaukee Harvester Company. They had agreed to a purchase price of $4 million—a great bargain—and could secure the deal with a $100,000 payment they had not yet made.

Perkins perked up at this information and offered that if they hired J. P. Morgan to form the harvester trust, Morgan would advance the $100,000 to hold the option to purchase the Milwaukee Harvester Company.

THE MERGER MOVEMENT

Between 1895 and 1904, more than one hundred large mergers occurred in America. The mergers typically formed what would now be consider monopolies, and they varied widely in structure and behavior. The harvester trust was a relatively benign union; others were predatory to consumers, investors, or both. *Library of Congress LC-USZC4-2835*

DISTRUST OF TRUSTS

The first of the large, high-profile American business mergers was Standard Oil in 1882, which led the state of New Jersey to pass an 1888 general incorporation law favorable to such unions. The company made John D. Rockefeller Sr. the richest man in America. By 1904, when this illustration was made, trusts and mergers generated significant public backlash. *J. Ottmann Lith. Co, Library of Congress, LC-USZC4-435*

J. P. MORGAN OFFICE BUILDING

At the corner of Broad and Wall Street in New York City, Morgan's headquarters were commissioned by Morgan and his first partner, Anthony Drexel, in 1872. The land alone cost $250,000—which the *New York Times* reported was the highest price paid on record for a New York City lot. Thomas Edison made history here, lighting six hundred of his brand-new light bulbs inside the building in 1882. The deal that formed the harvester merger occurred in this building in summer 1902. *Library of Congress LC-DIG-det-4a13508*

The meeting broke up around 1 p.m., and Perkins left the McCormicks to discuss. Before he could leave the building, the McCormicks ran him down and accepted the deal.[76]

This was June 20, 1902.

(As an aside, Nettie Fowler McCormick stated the decision to proceed with the Morgan firm was made on June 23, when a group of McCormicks met in Chicago at 135 Rush Street to discuss the issue.[77] She also said that she personally weighed the decision and held up the final contract an additional four days.)

By June 25, Middlekauf had secured the Milwaukee Harvester deal. Perkins was diving into the books of the McCormick company with Stanley in New York, and the Deerings had indicated their approval of the plan.

J. P. MORGAN

Drawing of Morgan as a young man. *Library of Congress LC-USZ62-78347*

On June 27, Stanley ran into John D. Rockefeller Jr. on Chicago's elevated commuter train going uptown. Upon hearing of the deal, Rockefeller Jr. told Stanley that his family would put up any necessary capital and instructed Stanley to inform Perkins of this without letting on where the money would come from.

To further reinforce this line of thinking, Harold met with John Rockefeller Sr. a few days later in Cleveland, where the interest was confirmed. Rockefeller Sr. added that if the merger failed, he would put up the money to purchase the Milwaukee Harvester Company. (Rockefeller Sr. was good for his word and would, in the coming years, invest as much as $60 million in the company.)[78]

Perkins, in the meantime, rapidly was putting together a deal that would assemble fierce rivals and proud leaders into a single company. He gathered the leaders in New York in July, wisely stashing them in separate hotels. By July 11, he had the McCormicks, Deerings, and the Plano leaders all in New York (the Milwaukee deal was done—and apparently with some hardball negotiation—only $3 million of stock was issued for it,[79] indicating a $1 million dollar price reduction).

James J. Glessner, the owner of Plano, would later claim that he was not aware that he sold his business to become part of a merger. The president of his company, Governor Bushnell, was invited to New York by Perkins, and Bushnell asked Glessner to join him in New York to meet with Perkins. Bushnell was quite ill at the time and had Glessner do the talking.

In the meeting, Perkins asked if they were willing to sell, provided the price was fair. With Bushnell's blessing, Glessner said they were. Negotiations continued and—according to Glessner—he learned of the merger on July 27, one day before he sold his company.

Perkins had to play his cards very carefully, as one of his key clauses was incredibly hard to sell. Rather than let any of the squabbling leaders rule this roost, the new company would be controlled by a voting trust. The members would be named at the sole discretion of Perkins, as official representative of the Morgan firm.

By July 11, Perkins had convinced the leaders in New York to accept these terms. And by the end of the month, the financing was agreed upon. The five companies would supply $50 million in cash and receivables, with J. P. Morgan supplying another $10 million in exchange for stock (at zero risk—Rockefeller Sr. and the McCormick family immediately offered to purchase the Morgan-held stock). A three-man committee would do appraisals of each of the five firms, and stock would be distributed according to the appraised value of each.

GEORGE W. PERKINS
The architect of the harvester merger was Perkins, a brilliant young financier known as J. P. Morgan's right-hand man. *Library of Congress LC-DIG-ggbain-07100*

The five companies that comprised the International Harvester Company were as follows, along with their valuation from 1903,[80] to give a relative idea of size:

McCormick Harvesting Machine Company ($26.3M)
Deering Harvester Company ($21.4M)
Warder, Bushnell & Glessner Co. ($3.4M)
Milwaukee Harvester Company ($3M)
Plano Manufacturing Company ($2.2M)

When formed in 1902, the harvester trust held about 85 percent of the total production of harvesting machinery in the United States. The company sold 90.9 percent of all binders, 81.2 percent of mowers, and 67.0 percent of rakes sold in 1902.[81]

JOHN D. ROCKEFELLER JR.
Harold McCormick and John D. Rockefeller Jr. were friends—this is how Harold met Edith. The two would encounter each other on the street and have a conversation that led to John D. Rockefeller Sr. helping finance the harvester merger. *Library of Congress LC-DIG-hec-06045*

NETTIE AND STANLEY McCORMICK (CIRCA 1900)
Both Nettie and Stanley would play roles in seeing the merger through. Nettie reported that during those fateful days in June and July of 1902, Stanley, Harold, and others stayed up all night working on the deal. She also stated that the stress of the negotiation wore greatly on Stanley. In 1909, Stanley was confined to psychiatric care at the family's Riven Rock mansion in Santa Barbara. The dramatic situation inspired the novel, *Riven Rock* by T. C. Boyle. *Cox Photography, Wisconsin Historical Society / 3650*

The Morgan firm—essentially Perkins—was charged with organizing the new company, incorporating it in the state of their choice, and appointing the Board of Directors and Officers. In the case of any disputes, the Morgan firm had final say. And, in fact, the Morgan firm would even name the company.

When Perkins cabled the Morgan firm with the details above on July 28, the response was enthusiastic: "HEARTIEST CONGRATULATIONS."

Perkins himself chose the name, International Harvester Company, wanting to ensure the name's appeal abroad as well as in the US. The company was incorporated in New Jersey, and the voting trust was made up of Cyrus Sr., Charles Deering, and Perkins.

The deal was signed on July 28, 1902. Astute historians will note that the founding date of the International Harvester Company is August 12, 1902. This is because the company's physical assets were all sold to a bank vice president, William C. Lane. Two weeks later, Lane sold all five companies to the International Harvester Company.[82]

(In court, Lane said that he was brought into the deal by a law firm and had no interest in purchasing farm equipment manufacturing companies. When asked whether he was paid for his role in the purchase, his response was: "Well, I have no recollection as to that." In fact, he uses that line forty-four times in his fairly brief court appearance. Given part of his sham purchase entailed making a $60 million payment, one might think the day would have made an impression on the man.)

Also note that only the company's physical assets were sold—not the value of their trademarks and such. All these machinations were done in an attempt to avoid anti-trust legislation. While this failed in the long term, it may have contributed to the fact that the harvester trust would not face legal issues for more than a decade. The fact that J. P. Morgan—unlike Rockefeller Sr.—was cozy with President Roosevelt also didn't hurt matters.

The arguments for the merger are an interesting blend of issues and certainly more complex than the overly-simple explanation that competition was too rough. The McCormicks may have been encouraged to merge because of the rapid ascension of Deering. While the McCormick operation was bigger—$5.4 million per year in 1902 versus $4.4 million that year for Deering[83]—the latter was growing much more rapidly than McCormick. And Deering was more profitable—McCormick had a 12 percent profit margin, while Deering clocked an impressive 18 percent margin.[84] William Deering was aging and tiring of the business, as was J. J. Glessner.

Global expansion was also a concern with the McCormick family. They had good success at the Paris Exposition of 1900 and were convinced there was more business to be had overseas if they could secure adequate funding.

All the harvester leaders cited the "reaper wars" as a tiresome issue, and certainly the extraordinary measures required to get a sale were a financial drain. Although they were all making pretty good money, with reasonable profit margins, prior to the merger.

Whatever their reasoning, in August 1902, the harvester trust found themselves with the lion's share of the agricultural equipment market at their disposal. Should the warring titans drop their gloves and work together, one would assume the profits would roll in.

As many know all too well, old disputes—imagined or real—do not just dry up and disappear with time. The harvester companies had a long and bitter history of litigation, competition, and dirty deeds, not to mention a massive network of dealers, suppliers, salespeople, factories, and suppliers strung out across a large country at a time when instantaneous communication was limited to telegrams.

Such differences would not fade easily or quickly, and the challenge of turning five large companies into one unified machine was greater than perhaps any of the founders imagined. As the founders jockeyed for position and struggled to make the five-headed dragon they had created behave as a unified beast, gearhead geniuses such as Legge and Steward would sequester themselves in basements, workshops, and warehouses, searching for the next mechanical stroke of brilliance that would transform the farm.

THIN ICE

Forming a merger in 1902 required great legal dexterity. The formation of the large companies that dominated market segments were regarded with suspicion by the public and drew the attention of the courts and lawmakers. The founders of the harvester merger structured the deal intricately to dance around the anti-trust legislation in place. *1903 illustration*, Puck *magazine*

EARLY MECHANIC
Wisconsin Historical Society / 161230

CHAPTER TWO

THE DAWN OF THE TRACTOR

SELF-PROPELLED FARM POWER EMERGES 1902–1922

"When farms are fully mechanized and tractor farming reaches its maximum efficiency, 15 per cent [sic] *of the population will grow all the food and fiber needed for our expanding millions . . . the key to all this progress is power."*
—A. C. Seyfarth, *History of a Tractor*

Like so many of the mechanical geniuses who turned technology into gold in early America, Edward A. Johnston began imagining and building new machines in his youth. By the time Johnston was eighteen years old, he had designed and patented a successful paper box-building machine. By the age of twenty, he had also produced the vertical-bar mower and, in his spare time, busied himself designing a phonograph, gas engine, and a machine that made X-ray photographs.[1]

When he joined the McCormick Harvesting Company in 1894,[2] Johnston's talents quickly made him a rising star in the McCormick Harvesting Company universe. He developed a steady stream of new technologies, including his crowning early achievement: creating and building the Auto-Mower that won first prize at the 1900 Paris Exposition.

The same year the leaders of the farm industry gathered in New York to form the "harvester trust," Johnston resigned over a patent dispute.[3] The loss was not an insignificant one for the company.

The merger would provide opportunities for growth, but that did not change the fact that fresh innovation was key for agricultural companies to thrive in this era. While much progress had been made in the nineteenth century, the true replacement of the horse on the farm would require internal combustion engines to provide power. The transformation was a hotbed of opportunity for those with the genetic

TITAN ENGINE

The earliest gasoline engines to provide farm power were portables on carts, like this Titan engine shown in 1906. *Wisconsin Historical Society / 41501* predisposition to visualize complex systems and the boots-to-ground, dirt-under-their-fingernails grit to forge their visions with iron, oil, and wood.

The leaders of the newly created International Harvester Company (IH) had to be pained at the loss of Johnston in 1902, but much of the company's energy was focused on sheer survival.

The merger was hardly the boon one might expect. In fact, while it was quite profitable for the company founders, the immediate effect on business and profits was less than satisfactory. The massive headache of merging fierce competitors with a long history of bitter rivalry, during an era when communication technology was limited, would take more than a decade to sort out. The leaders bickered over low profits, valuations, and how to divide the markets. The five companies largely continued to operate like five entirely separate organizations.

Part of this was strategy—the leaders determined that they were better served to continue marketing their independent brands—and part was due to the massive endeavor of combining such large entities.

Far from the vitriol brewing in boardrooms in Chicago and Plano, new technology was simmering in blacksmith shops, converted carriage houses, basements, and warehouses across the country.

The most exciting new technology on the horizon was the tractor. Company engineers like Johnston and John F. Steward, as well as the ever-resourceful well of farmyard tinkerers, were creating an array of fantastic machines worthy of a

EDWARD A. JOHNSTON

Johnston was one of the star engineers of McCormick Harvesting Company and IH. In a remarkably productive career, he would receive more than 171 patents, making him one of the most prolific agricultural engineers in recorded history. He was the man behind the McCormick Auto-Mower, the Auto Buggy, the Mogul tractor line, and he would go on to lead the IH engineering team. *Wisconsin Historical Society / 35532*

GEORGE ELLIS EXPERIMENTAL CAR

One of the early engineering pioneers at Deering and IH was George Ellis. He is shown with an experimental automobile, most likely in John F. Steward's basement. *Wisconsin Historical Society / 58395*

CASTLES TRACTOR

This extremely large experimental tractor was built in Australia, circa 1915. *Museums Victoria Collections*

IH EXPERIMENTAL AT DEERING WORKS

This photo, dated March 26, 1919, is identified only as a model "built and never used." Edward A. Johnston hand built a car in 1898; Alexander Legge had a soft spot for buggy manufacturers and built a powered buggy. This could be either of those—or something else entirely. Given Deering Works was Steward and Ellis's home base, it's also possible one of them created it. *Wisconsin Historical Society / 24877*

steampunk comic book. Farm magazine advertisements of the day showcased mechanical creations straight out of a Jules Verne novel. Part whimsy and part innovation, the ads were placed by innovative builders, cutting-edge entrepreneurs, and a handful of scheming scammers.

Engine power first came to the farm in a stationary form. The first of these were steam engines, which provided belt power for harvesters and other large equipment. These machines were expensive and difficult to transport. Some farmers formed co-ops to buy and share this equipment, while others paid traveling businessmen for use of the equipment once or twice a year.

Steam engines were heavy, and transporting the early portables could require a team of horses. The first necessary technological leap was to use the big engine to power itself. Innovation on this front, at least as far as recorded history goes, took place in the Northeast. The first recorded steam plowing event happened in Pennsylvania. J. W. Fawkes built a plowing drag in 1858 and used a self-propelled steam engine to pull eight plow bottoms at 3 miles per hour. Others in Connecticut, New York, and New Jersey were busy building and patenting similar machines.[4]

The steam machines proved helpful for powering large farm equipment such as threshing machines and useful for opening large tracts of raw prairie, but these lumbering, complex beasts were not the solution for the average farm. As gasoline engines became more viable in the latter 1800s, sharp-eyed innovators were drawn to them as a more elegant solution for motive power. They were lighter, didn't require the time-consuming process of firing a boiler, and burned the widely available and incredibly cheap gasoline. The engines were also generally smaller than steam engines and easier to start and maintain.

The next logical step was a gas engine for farm use. Even the small ones were fairly heavy, so they were mounted on carts that could be drawn with horses. As larger engines were created, they became too heavy for even horses to pull . . . so why not have this large engine provide its own motive power?

Gas engines would be the key to the tractor, and their ease of starting would displace steam. This would also allow for the creation of small machines—machines that would ultimately be able to replace the horse. As with other innovations, with the technology available, a variety of people recognized the advantages of the gasoline engine invented by Nicolaus Otto. His innovation was the four-stroke cycle, which brought in fuel, converted it to power, and expelled the exhaust gas. He produced several variants, and the most famous and successful variant was completed in 1876.

As was evident with the reaper, innovation is generally not the result of a single stroke of genius, but the product of small advances made by a variety of people who are often interconnected. Partial solutions are built, innovators see what others are doing, and they build machines that advance technology. Such was the case with the farm tractor.

In a fascinating aside, IH employee and researcher A. C. Seyfarth wrote in his seminal work, *History of a Tractor*, "The farm tractor, unlike the reaper, cannot be credited to the efforts of any one individual." The irony of course being, McCormick was not the sole inventor—the reaper was also developed with contributions by a wide variety of innovators.

Rhetoric aside, the history books and, well, humanity love to point to someone as the first. And while it is often true that most innovations arise from a number of different inventors building on each other's work, we can examine the records and come up with a single person who the records show was the first. This begs the question: who built the first gasoline-powered farm tractor?

COMPLICATED ROOTS
In the early days of the International Harvester Company, advertising continued to show the merged companies as independent entities. This was a conscious choice, and the megalithic "harvester trust" was slow to turn its market dominance into efficiency. Osborne Works, shown here, was part of a 1903 acquisition and would later become known as Auburn Works. *Wisconsin Historical Society / 39505*

An individual who is widely credited with creating the first gasoline tractor was one of those backyard tinkerers, John Froelich. A large-scale farm owner based in Iowa, Froelich mated a gasoline engine to a home-built frame and drivetrain in 1892. While he would die penniless, his legacy has been preserved. His machine lives on in a museum in a town named after him, and the Froelich tractor has been widely lauded as the first gasoline-powered farm tractor.

The fascinating twist to that story is Froelich encountered another gasoline tractor, one more finished than his own rather crude creation . . . and running *four years* before he built his machine.

That machine Froelich encountered was the tractor designed by Franz Burger of the Charter company, and the story of it changes the broadly understood history of tractor engineering.

FRANZ BURGER'S ENGINE

The spark for the gasoline tractor was lit in Philadelphia in 1876, when the city hosted the Centennial International Exhibition, the first World's Fair held in the United States and a celebration of the one hundredth anniversary of the signing of the Declaration of Independence.

Always a showcase for new technology and a draw for the innovative geniuses of the era, one of the stars of the 1876 World's Fair was the Otto four-cycle gasoline engine.

One of the many people inspired by the engine was John Charter, a business leader whose visionary talents were offset by the fact that he was an opinionated and quarrelsome schemer. He traveled with George Robinson, his hard-working, loyal, and underpaid office manager.

At the Philadelphia Exposition, the Otto gasoline engine struck Charter's fancy. Charter had several jobs—he ran a combination foundry and machine shop, the Williams & Orton Company. He also was president of the local town gas company.

Town gas was derived from coal and piped in to supply heat and lighting. It would be replaced by natural gas, but not until the 1950s. In the early part of the century, coal-produced town gas was the household and business fuel of choice.

The Otto four-cycle engine burned town gas—which was in gaseous not liquid form—and this had obvious business appeal to Charter.

Down the street from the Williams & Orton office, a Mr. Elliot had a grain elevator that he powered with a steam

CHARTER GAS ENGINE CO.

The Charter headquarters in Sterling, Illinois. *Sterling Rock Falls Historical Society* engine that took up to an hour to fire up. Ever the salesman, Charter convinced Elliot to purchase an Otto four-cycle engine for the elevator. Apparently struck with a vision, Charter took his team of mechanics over to the mill to see the engine in operation and told them this was the very unit they were going to build “by the hundreds.”[5]

Charter was sharp enough to understand his vision required technical expertise, and he cleverly turned to patent records to find the mechanical innovator he required to realize his dream. His patent record research led him to travel to Washington, DC, to call upon Franz Burger. A skilled German mechanic, Burger was half-owner of the Matz & Burger shop[6] in which he made patent drawings and constructed the scale models often required by the US Patent Office.

Burger had designed an engine that burned liquid fuel (rather than vaporous town gas) in 1872–1873. Burger received a patent for his design in 1878. Charter found the patent and sought out Burger, visiting him in early 1882 to ask him to come to Sterling, Illinois, and build an engine for manufacture.

Charter made a host of promises to Burger, which Burger found so enticing that he left the shop in the hands of his partner without taking a single penny in return.

One problem with the new engine was getting around the Otto patent. Burger did so by making his gas-burning engine a two-cycle rather than the Otto four-cycle. Another hurdle to overcome was vaporizing liquid fuel. Burger developed a spray nozzle that did that—and this was his key invention.

A few 6-horsepower Burger gasoline engines were built and sold in 1884 and 1885, with satisfactory results. Burger personally believed that the future of the engine lay not in burning heavy oils, but lighter and cheaper gasoline. Charter had no interest in this direction and vehemently opposed it. Burger’s key work was the development of the spray nozzle that would safely inject the fuel into the engine’s cylinder using a needle valve. He did so under considerable pressure to stop.

“It often seemed a wonder to me that my father continued with this work on account of the opposition that he had,” Lee Burger wrote. “Somehow he stuck with it; there a perfect working engine was developed about 1886 or 1887, and sold to Lawrence Brothers Hardware Co., Sterling, Ill.”[7]

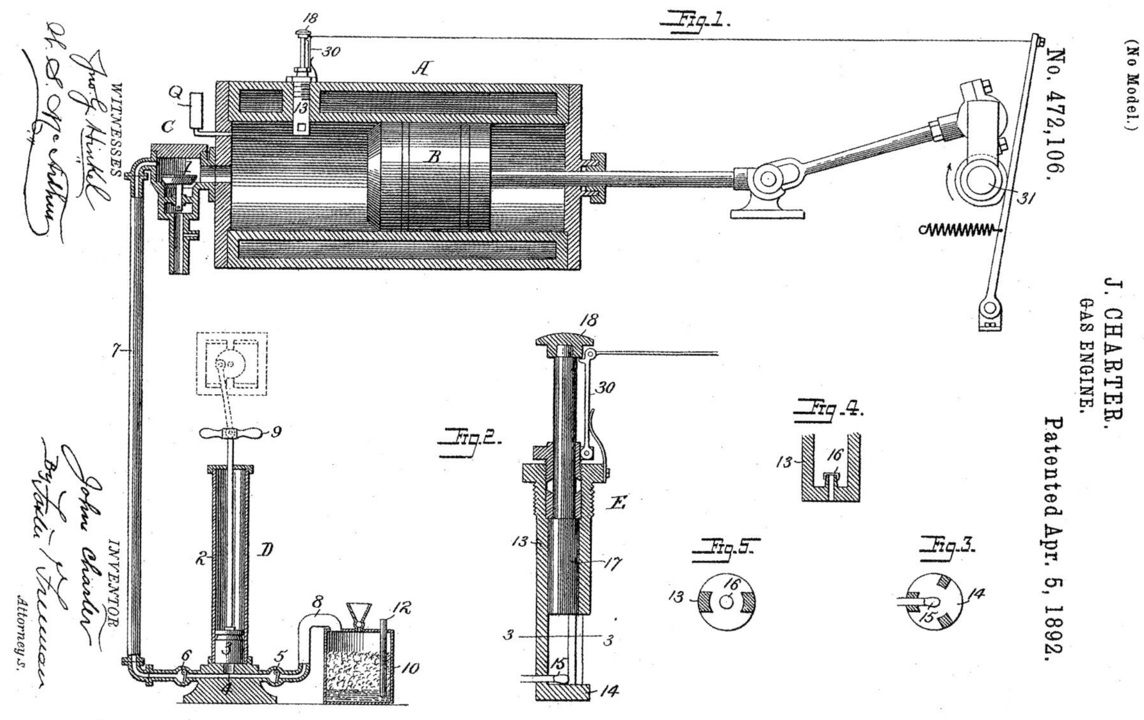

JOHN CHARTER PATENT
Franz Burger designed this key component for vaporizing fuel in a gasoline engine. His employer at the time, John Charter, filed for the patent.
US Patent Office

Despite his opposition to the creation of the engine, Charter filed for a patent on Burger’s fuel-vaporizing device, which reportedly earned him thousands of dollars per year in royalties from others who used it on gasoline engines.[8]

In 1889, the hard-working Robinson was rewarded for his dedication by replacing the irascible Charter as the leader of the Williams & Orton Company. In a fascinating turn, the company also changed its name to the Charter Gas Engine Company.

Infighting amongst leadership ensued, slowing sales, and Charter was eventually ousted from the company. He replicated the engine at the H. W. Caldwell Company in Chicago but became frustrated and sold the design to Fairbanks-Morse. John’s son, James Charter, eventually became chief engineer at Fairbanks-Morse.

Burger’s engine would, however, lead to creation of the first farm tractor—which would eventually be swallowed by the sprawling corporate monolith of the agricultural industry, International Harvester.

KEYSTONE AND THE BURGER TRACTOR

By Guy Fay

By 1886, the Charter Gas Engine Company was selling gas engines, and it may have been the first company in the United States to build and sell portable gas engines (which were simply engines mounted on wheeled carts, typically pulled by horses). Shortly thereafter, independent of Charter Gas Engine, John Charter and Franz Burger began experimenting with a self-propelled gas engine—a machine then known as a gas traction engine, now known as a tractor. Their work began on behest of another company based in Illinois—Keystone Manufacturing.

Keystone was founded by Thomas A. Galt in 1857 in Sterling, Illinois. After a fire in 1867, the company moved across the Rock River to Rock Falls. Keystone built farm implements, including one of the first husker shredders, a line of corn shellers, as well as tillage and planting implements.

BURGER TRACTOR

This photo is the only surviving record of what is likely the first version of the tractor that Keystone and Charter were developing, identified by Lee Burger as being worked on by his father, Franz Burger. The tractor may have used a wagon to carry sufficient cooling water; one report has the tractor working with a water tank on a separate wagon for belt work and relying on a small cooling tank for moving between farms. *Wisconsin Historical Society / 22881*

Keystone used "horsepowers" or "horse sweeps," to power large implements with horses. The horses are hitched to a unit that looks like a merry-go-round. The horses walk in a circle, turning the horse sweep, and the horse sweep transfers power to the implement through a rotating shaft. Similar machines were used by other companies such as J. I. Case to power threshing machines and other equipment.

Unlike the threshing machine companies, the Keystone company did not manufacture portable steam engines that could power even larger machines or steam traction engines that could move the rigs from place to place. Keystone's advertising portrayed its largest shellers and shredders paired with generic steam engines. Keystone ads showed the smaller machines being pulled from farm to farm by some very hard-working horses.

At some point shortly after the Charter gas engine started selling well, Thomas Galt and John Tracey of Keystone made a deal. They wanted to produce the Charter gas engine for the European market, and they purchased a half interest in the European gas engine patents owned by John Charter. On March 2, 1887, the *Sterling Daily Gazette* reported John Charter was building an engine specifically for the American Manufacturer's and Inventors Exhibition in London, with his son scheduled to travel there to run the booth, which was to be manned strictly by Americans.

John Charter also designed and built a gas tractor for Keystone, reportedly for $10,000. Using a 20-horsepower Charter gas engine, the Keystone gasoline traction engine was tested powering a six-hole corn sheller on the farm of Sheldon and Stone. The September 13, 1888, issue of the *Sterling Daily Gazette* reported it shelled 850 bushels of grain in four hours with seven-and-a-half gallons of gas at ten cents per gallon. Galt and Tracey were reported to be in the process of making "a few slight changes" and getting the machine into production. Author C. H. Wendel identifies this tractor as one of the first gasoline tractors built.

The Keystone gas traction engine history more or less ends there. One report said that it didn't work well while pulling a load on rough ground.[9] Franz Burger's son, Lee, was sent to be an expert on the gas traction engine to try to work out its issues. As an aside, Lee Burger would eventually take a job at IH, where his work as an engineer was critical to creating IH's new diesel engine in the 1920s.

CHARTER / STERLING TRACTOR
In the mid-1890s, Charter resumed the tractor project. This machine, which was often referred to as the Sterline Tractor, has a rudimentary cooling system, although it may have relied on a separate cooling tank. With the huge belt pulley, this is obviously a low-rpm engine.
Sterling Rock Falls Historical Society

The Keystone company would continue making a variety of agricultural implements and would be purchased by the International Harvester Company in fall 1904.

As for the tractor, Thomas Galt was reported to still be working on it in 1892, but after that, Keystone faded from the story of the Burger tractor. An 1893 lawsuit seemed to blow up the relationship between John Charter and Charter Gas Engine Company, and this may have caused the withdrawal of Galt and Tracey. Charter Gas Engine Company made a short run of redesigned gasoline traction engines in 1894–1895 and sold them in South Dakota.

A young South Dakota thresherman who started with steam in 1888 purchased either a Charter stationary or portable gas engine to learn the new technology, and following the Keystone/Charter gas traction engine, he decided to build his own. For some reason, possibly due to the issues with the Charter, he would purchase a Van Duzen vertical gas engine to put on his chassis. His name was John Froelich.

CHARTER / STERLING TRACTOR
With its narrow rear wheels, low horsepower, and rudimentary cooling system, the Charter could move to machinery on good roads and farmyards but had issues in softer or rougher ground. *Sterling Rock Falls Historical Society*

While the Froelich tractor built in 1892 is widely credited as the first gas traction engine—or tractor, if you will—the Burger tractor predates that machine by four years. Given Froelich's purchase of a Charter machine, it's highly likely that the inspiration for his machine came from the Burger tractor.

The first gasoline-powered farm tractor known to exist is the Burger tractor, designed by Franz Burger, financed by John Charter, and first tested in 1888.

McCORMICK AUTO-MOWER

Edward A. Johnston on one of his two Auto-Mowers, probably the single cylinder. The photo was taken in the extensive lumberyards next to McCormick Works. A surviving Auto-Mower resides at the Wisconsin Historical Society in Cassville. *McCormick Harvesting Machine Company, Wisconsin Historical Society / 39487*

THE SECRET START AND PUBLIC LIFE OF THE AUTOMOWER

By Guy Fay

Keystone may have been the first IH predecessor to experiment with gasoline traction engines, but not by much. George Ellis, hired by Deering to experiment with making binder twine with US materials, also had an interest in gasoline engines. Ellis began experimenting with gas traction engines as early as 1889, shortly after starting at Deering.

At some point, William Deering had moved gas engine experimentation into John F. Steward's basement. Basement experimentation was intended to keep development a secret, or at least to keep it away from prying eyes. Despite the desire for stealth, there were plenty of visitors to Steward's basement, including one Henry Ford.[10] And when they got to the stage of operational models of self-propelled machinery, as they eventually did, they had to take the machines out of the basement.

One of the early machines they were constructing was the Automower. They started testing different variations in fields at least by 1891. These were small machines. One of the innovations on the machine was that the engine rather than the wheels was used to drive the implements attached to the Automower—typically Deering grain and corn binders. A hitch was added, and horizontal engines were used on Ellis's design.

The leaders of the Deering Harvesting Machine company decided to take the Automower to France for display at the 1900 Paris Exposition. The competition between Deering and McCormick was fierce and global at this time, and both companies were planning large displays of current machines as well as gigantic displays of historical machines and models to support each side's claims of inventions. McCormick caught wind of Deering's plans to display its Automower, which spurred the McCormick side into action.

McCormick called on Edward A. Johnston to produce an answer to Deering's Automower. Johnston was already an experienced harvesting machine inventor at the age of eighteen. He had been designing new harvesting machines for McCormick but had also been developing automobiles as a personal project. He was given very little time to produce a competitive automower for the showcase in France.

Johnston's quickly created McCormick Auto-Mower delivered in style and implemented some new ideas not found on the Deering concept. Johnston made the engine directly run the mower bar separately from the drive—meaning it functioned similarly to what would presently be called a live power take-off. In fact, he ended up building two Auto-Mower models—one powered by a single-cylinder engine and another driven by a twin-cylinder. The single still exists today and was created first. It was built quickly and displayed in operation in France before the Jury of Awards, but it apparently didn't do so well (according to Steward at least, who was in charge of the Deering display at the Paris Exposition). The McCormick two-cylinder was built later and was powered by a vertical twin.

When the two-cylinder McCormick Auto-Mower was taken to a heavy field of French alfalfa to compete with the

Deering Automower, the direct-drive mower netted a McCormick victory due to better performance in heavy crop. The demonstration attracted an enormous amount of attention, including "the crowned heads of Europe" and Nettie and Cyrus McCormick Jr. The outcome was calamitous, violence was threatened, but it was averted as the Deering folks let out their frustration by trashing the generously provisioned McCormick hospitality tent. In *The Reaper*, Steward claims that the McCormick and Deering "had about equal honors."

The Deering Automower was tested and displayed until at least 1906, including at the St. Louis World's Fair of 1904. The Deering Automower played a part in a political campaign in the early 1900s, generating free publicity.[11] Newspaper commentators pointed out that while automowers promised a wonderful future in internal combustion for the farmer, they were heavy, expensive, and pretty much single purpose, which was probably why they were never sold.

Both the Deering and McCormick automowers influenced tractor development. The Deering engine was part of the evolution that would become the Deering and then International horizontal stationary engine introduced in June 1904. This engine put IH in the internal combustion business to stay, and in a big way.

Both automowers captured a lot of attention, particularly due to their ability to directly power a cutting bar. Albert Gougis, a French implement manufacturer, built his own small tractor that could power an implement. He added an important twist: he powered it with a rotating shaft strung between the tractor and implement. While the rotating shaft had been used in steam traction engines and a few large gas tractors to drive a thresher, it was in stationary use and had never really replaced the belt.

Gougis built a shaft with two universal joints that would continue working with the tractor both in and out of gear. It was used to power a grain binder.

Before the Gougis machine disappeared into obscurity, a few IH employees observed it in action. One of the IH people who saw the machine, and was likely influenced by it, was Edward A. Johnston. He would become IH's chief engineer and would oversee the development of the International 8-16, which was modified a few years later with a similar rotating shaft arrangement—the power take-off (PTO)!

DEERING AUTOMOWER

The Deering Automowers went through several generations of development in the 1890s and 1900s. While they were never produced for sale, the engines were developed into the International line of stationary engines by George Ellis, who is probably the driver here. *Wisconsin Historical Society / 27695*

DEERING AUTOMOWER

The Deerings were displayed many times over the years, with plenty of publicity! *Wisconsin Historical Society / 60675*

STRUGGLES OF A NEW CORPORATION

When International Harvester formed in 1902, it was gifted with much more than the development work done on a few small gas-powered machines. The founders had created a megalith that dominated the agricultural industry. Making that giant monopoly function properly was nearly as formidable a task as creating the first successful gasoline tractor. In fact, the process of bringing together five warring leaders resulted in internal political battles, saw the individual firms continue to fight each other for sales, and was an accountant's worst nightmare.

In the first few years of the company, the difficulties of managing the sprawling organization along with disagreements between the owners undermined International Harvester's ability to transform market dominance into profit. Profits were below expectations of the shareholders as well as the owners.

LOOK THE OTHER WAY

President Teddy Roosevelt evinced himself as a trade buster, but the truth was he was willing to look the other way. International Harvester would benefit from this, particularly in 1908, when the Roosevelt administration chose to ignore a report recommending the company be prosecuted for antitrust violations. *1904 illustration,* Puck *magazine*

A key factor of the merger was that the companies were controlled by a trust, and, perhaps more significantly, George W. Perkins had the power to decide who was in the trust and would cast the deciding vote in the event of stalemated decisions. He later described this as "unfettered control" of the organization. Interestingly, Stanley McCormick said that when the Deerings became aware that John D. Rockefeller was investing in the new company, they ended up "favoring, if not insisting upon, a voting trust."[12]

The Perkins control of the new organization reflected experience—he and J. P. Morgan had learned much managing the foundation of US Steel, which formed as the world's first billion-dollar company in March 1901. As complex as it was to put together such entities, the "most difficult problem facing [US Steel] was the reconciliation of internal differences."[13]

Their difficulties managing warring leaders during and after the steel merger helped them understand that they would need to carefully manage the leaders of the five companies in the harvester merger—particularly the Deerings and McCormicks, who had a long history of clashes. Perkins wisely crafted the trust as an effective tool to mitigate the inevitable battles.

Cyrus McCormick III alluded to this when he wrote, "Perkins had sensed that no McCormick and no Deering could long remain at peace with each other . . . The voting trust served to tide the new company over a difficult trial period."[14]

The board of directors Perkins organized to control the trust was carefully selected, with five members from the McCormick organization, four from Deering, and five others (including Perkins) loyal to the Morgan concerns (one of those being Elbert H. Gary, the founder of US Steel). This balance gave Perkins and the Morgan firm strong control. He could serve as a peacemaker, parent, or despot—whatever the situation required.

Early in the merger, Perkins started off on the wrong foot when he suggested IH's steel and ore assets be sold to US Steel, along with a ten-year contract locking IH in to purchasing all its steel from US Steel. This would have been a significant loss of cost control for IH.

The Deerings and the McCormicks refused the offer. Both were not happy with the proposed interference in their company, especially given the proposal was favorable to another

Left:

HAROLD McCORMICK

The younger brother of Cyrus Jr., Harold would play leadership roles in early IH and step up as company president in 1912. His marriage to Edith Fowler McCormick would link the family to John D. Rockefeller, who would help fund and create IH. *George Bain Collection, Library of Congress LC-DIG-ggbain-07872*

Right:

CHARLES DEERING

Deering company founder William Deering retired in 1901 and turned over the company to his sons, Charles and James. The two brothers worked with the McCormicks to found IH, and Charles became chairman of the board. He retired in 1910 and devoted his life to collecting art. *University of Miami Library, Special Collections*

Perkins and JP Morgan client but perhaps not in the best interest of IH.

Perkins would later acknowledge his move was a mistake—one he did not repeat.

THE PROFIT PROBLEM

Profits were a problem in the early years. They were so low that most were returned to investors in the forms of dividends (see chart in appendix). Perkins wanted to release more stock, and the McCormicks stridently objected to that, as it would increase dividend payments to stockholders. As it was, the trust had expected to pay 6 percent dividends when they formed—they paid only 3 percent in 1903.

There is some question if the low profits shown in the early years of IH are accurate. The five large companies didn't come together easily, and consolidating the books was a complex and difficult task, so accounting errors were a real possibility. Also, high profits would not help IH avoid antitrust legislation, nor would high earnings figures play well with the public. So, it was in the company's best interest to keep the numbers low. In an interesting footnote to this, a September 1902 memo informing Champion dealers they were now part of IH included the order to pay agents in cash.[15] Certainly this would make accounting more difficult. Cash payments would also make it easier to manipulate the figures.

The US Bureau of Corporations (BOC) would find that discerning the fiscals of IH in those very early days was a confusing morass—to the point that "the accounts were for several years kept in an extraordinarily loose manner" and "the company, according to the statements of its comptroller, had actually made no complete and authentic balance sheets prior to that for December 31, 1906."[16]

Due to the way the records were kept, the bottom-line truth of how much money the company made in the earliest days is lost to history.

BRAND INDEPENDENCE

One of the first decisions made at the newly formed harvester trust (IH) was to keep all its branding intact. This decision was made at a meeting in August 1902,[17] and the business continued to operate that way for well over a decade.

The decision to operate the brands independently meant very little changed for the customers, who were presented with choices very similar to what they had prior to 1902.

GEORGE PERKINS

The architect of the merger that formed IH became the puppet master. Perkins (left center) carefully crafted the structure of the voting trust that ruled IH from 1902 to 1912 and held the right to call the shots when the trust couldn't come to an agreement. *Library of Congress LC-DIG-ggbain-21961*

The dealerships remained intact as well, meaning some cities would have four or five seemingly competing dealerships that were all owned by IH. Oddly enough, perhaps because IH feared an antitrust suit, it did not use its power to raise prices until 1910, which was part of the reason profits remained low.

Ultimately, the brand confusion it propagated cost it market efficiency. All the contortions and fiscal chess moves only delayed matters, as the government eventually decided to file an antitrust lawsuit against International Harvester anyway.

The founding members of IH, along with George W. Perkins of the J. P. Morgan firm, performed a number of shenanigans and stunts to attempt to avoid antitrust litigation. The acquisition of Milwaukee being the most egregious, with the purchase of Milwaukee done secretly by a third party—Mr. P. D. Middlekauf—to prevent the sellers from knowing who purchased them.[18] The records are so murky that, in 1913, the courts could not accurately ascertain whether the purchaser was J. P. Morgan, George W. Perkins, or the McCormick family. Also, the $100,000 advance payment was kept off the books at McCormick as part of the deal.

Additionally, all five of the IH founding companies were initially purchased by a middle man, William C. Lane, who held them for several weeks and then sold to the newly formed IH.

EARLY IH ACQUISITIONS

Almost immediately after its founding, IH began to purchase other companies. The subterfuge of the merger continued in these negotiations. D. M. Osborne Co. was purchased in January 1903. The purchase was done by officers of IH rather than the company itself, which appears a rather clumsy attempt to mask the purchase.

In 1904, it purchased the Keystone Co. in the same manner. The acquisition was kept more or less a secret, and Keystone and Osborne advertised themselves as independent long after the sale to IH was complete, which extends well past clumsy maneuvers and into the unethical (and the US government would use the false advertising as a component of their case against IH, particularly in the 1913 summary report).

IH soon purchased the Minnie Harvester Co. of St. Paul, Minnesota, and Aultman-Miller Co. of Akron, Ohio. In 1904 and 1905, it added wagons through the purchase of the Weber Wagon Co. of Chicago and began purchasing steel frames and wheels from Bettendorf (see appendices for charts showing acquisitions and subsidiaries 1902–1922).

The expansion was alarming to the competition. While there had certainly been enough of that during the reaper wars, the agricultural companies often used a surprisingly gentlemanly tactic of sticking to certain territories or lines and writing letters notifying each other of intent.[19]

One of the largest competitors, Deere & Co., had initially responded to IH with cautious deference. After all, IH made harvesters, while Deere had plows and cultivators. They also simply were outgunned in this era: Deere's valuation was just under $4 million in 1900, and IH was initially established at $120 million. Deere had gross sales of $2.1 million in 1899–1900, and IH's gross sales in 1903 alone were $52.0 million.

Deere & Co.'s somewhat timid approach to IH in that era was more the wisdom of a mouse avoiding a large, hungry snake rather than an overabundance of corporate caution. Even given the disparity, Deere's optimism dissipated as IH began to rapidly expand into new markets. A few Deere & Co. executives, particularly Charles C. Webber, pushed Deere to expand into IH-dominated product lines—particularly harvesters. The easy way into that market was to purchase Acme Co., which IH was also pursuing. In the end, Acme was found wanting by both purchasers, and Charles Deere didn't have the stomach for Weber's proposal.[20]

In 1910, IH went head-to-head with Deere's best market when it expanded into plows, contracting Oliver Chilled Plow Co. to supply plows for the company to sell. IH upped the ante in 1919, purchasing Parlin & Orendorff Co. of Canton, Illinois.[21]

IH would also purchase J. S. Kemp Co., the American Seeding Machine Co., as well as hardwood timber lands. It also attempted to purchase Massey-Harris in 1902 but was not able to come terms.

While managing the company was a challenge, IH certainly used its financing and market power to add significantly to its market stranglehold.

SHIFTING CONTROL

As IH leveraged its position to increase its market share in the agricultural industry, the McCormicks were increasing their ownership and control within the company. Over time, the Deerings would steadily decrease control and ownership in the business. This isn't entirely due to the McCormicks squeezing them out; investors often have different goals and appetites for risk, and the Deerings stated on several occasions their interest in exiting the business with compensation.

The Deerings desire to get their money out led to a significant reorganization made effective October 29, 1906. The most notable portion of the agreement took the company public (offering stock for sale on the stock exchange) in 1908. This allowed the Deerings to trade their stock position for cash.

The agreement also included dozens of minor clauses that changed the balance of power, tilting slightly in the favor of the McCormicks. Ultimately, this would work out well for IH, as profit margins finally began to improve, particularly after IH raised prices and expanded into more international markets in 1910.[22]

As time wore on, the Deering family would see their role in IH leadership reduced. The McCormick family, on the other hand, would hold on to (and even augment) their legacy after the merger and become a guiding influence directing the course of IH.

In fact, IH would have McCormick family leaders pretty much continuously into the 1980s. While Fowler's moves in the 1950s would prove overly ambitious and labor relations were an issue throughout the company history, the McCormicks deserve significant credit for helping IH survive the difficulties of the merger while using their stranglehold on the market in relatively responsible ways and going on to build one of the most storied legacies in farm equipment manufacture.

MAKING HAY

Particularly prior to 1908, IH carefully kept all its multitudes of brands separate. With Osborne, a 1903 acquisition, IH continued to advertise them as "independent" for several years. This is a 1906 catalog cover. *Wisconsin Historical Society / 29140*

KEYSTONE AFTER THE IH MERGER

By Guy Fay

By 1903, Keystone Manufacturing had been insolvent for several years.[23] Some former local McCormick company employees, led by Harold B. Utley, bought the Keystone company in March 1903.[24] He and his investors attempted to revive the company, at least to the point where they could make some quick money. Keystone advertised that it was the David to International Harvester's Goliath, and quite a few IH men moved to Keystone's employ.[25]

The company started adding grain binders and mowers, especially after Ed Johnston left IH for Keystone.[26] Keystone had a rich history of corn planters, shellers, and other equipment. Its new mowers and binders moved the company into direct competition with IH, which virtually owned the binder and mower business. In 1903, IH sold 98.15 percent of the binders sold in the US and 92.05 percent of the mowers.[27] Johnston's new mowers were a real improvement, and IH had to be nervous about competition from a man they knew was a formidable generator of innovative machines.

On October 13, 1904, Keystone sold a controlling interest in the company to a group of officers, stockholders, and directors of the International Harvester Company. This group included Cyrus H. McCormick, Charles Deering, and George W. Perkins—meaning the sale was effectively to IH.[28]

In September 1905, IH consolidated the purchase,[29] resulting in accusations about monopoly violations.[30] Rumors had started in July 1905,[31] especially when an IH Executive, R. B. Swift of the IH Patent Department, was indicted by a grand jury for embezzling large amounts of funds from both the old McCormick Company and the new International Harvester by buying patents (some useless) for more than what they were worth and then taking kickbacks.[32] Swift had been one of the IH executives involved in the purchase of Keystone in 1903 and then approached Cyrus McCormick with the proposal to sell Keystone to IH in August 1904.[33] Swift took revenge on IH, including confirming the Keystone deal. The implement newspapers printed accusatory articles,[34] and salesmen who had left IH were printing their accusations of betrayal.

IH soon discovered that they had seriously overpaid former IH employees for the company, maybe as much as double what it was worth. But they did bring some former IH employees

KEYSTONE ADVERTISEMENT 1904

Keystone played up the David versus Goliath theme for a few months not only to customers, but to potential dealers and employees who were suspicious of the new IH Company. *Public Domain* back into the fold, including Johnston, who had been busy in the meantime with internal combustion experiments as well as with the grain binder and mower.

The grain binder was a chief attraction to the IH top management. Johnston had made a much lighter binder, which McCormick and IH had been attempting to do for quite some time.

In addition, he had shipped several automobiles he had built to his new home in Rock Falls. While at Rock Falls, either when employed by Keystone or by IH, he had started work on a larger tractor, which in later documents he vaguely described as the start of the Mogul developments and possibly the tractor that went to the Winnipeg competition in 1908. He also was involved in developing the first Auto Buggy while in Rock Falls.[35]

All of this harkens back to the automower, an odd little machine that was never sold that has nonetheless become a cornerstone machine in IH's development of the tractor.

KEYSTONE BROCHURE 1913

Keystone survived as a separate brand of IH for many years, filling in certain holes in IH's efforts to be a "long line" agricultural equipment company, with goods in demand all year, not just during harvesting season. *Wisconsin Historical Society / 23320*

KEYSTONE
HORSE-POWER

KEYSTONE LINE

Keystone had a line of implements and machines suited to the local Northern Illinois market, especially corn, but it sold implements all over the Midwest and occasionally elsewhere. Some tillage was included, but not plows, where there was enormous competition. *Guy Fay Collection / 1892 Keystone Catalog*

1899 S. S. MORTON TRACTOR

This machine was built by Samuel S. Morton in May 1899 and used a friction drive. In a note attached to this photograph, Morton wrote Morton Manufacturing Company sold fifty-five to sixty machines of this type. *Wisconsin Historical Society / 90990*

THE FIRST INTERNATIONAL GAS TRACTORS

When IH purchased the Keystone company in 1904, it returned its star engineer, Edward A. Johnston, to the fold. The natural-born gearhead continued to develop gas engines as well as a three-wheel tractor[36] at Keystone Works in Rock Falls, Illinois.[37]

While these were the early days of tractors, Johnston was far from the only innovative engineer developing new self-power machines using the rapidly evolving gasoline engines of the time. IH's steadily growing team of engineers had created a new line of gasoline engines that were selling well, and the farm tractor was attracting the attention of IH leadership as well as the engineering departments of several IH divisions.

The tractor was just emerging when IH formed. The first successful production gasoline tractor would come out of a company founded by two young University of Wisconsin students, Charles Hart and Charles Parr. Their Hart-Parr Company sold its first tractor in 1902. Despite this machine destroying a rickety wooden bridge and dropping into a muddy creek while being delivered, it was pulled out by horses, repaired, served seventeen good years with the new owner, and was scrapped shortly after World War I.

By 1904, the two mechanically-oriented honors students had built the Hart-Parr Company into a dominant force in the nascent market, holding roughly half the tractor business.[38] Bear in mind this entailed a few hundred machines at best—tractors in this era were large, expensive specialty items for the biggest and wealthiest farms and contracting outfits. Hart-Parr was a pioneer in the business, credited with building the first successful production tractor, and the largest of the dozen or so companies who were building tractors at that time.[39]

By 1905, a number of manufacturers were beginning to see the promise of the tractor, among them IH, which decided to enter the market by mating a purchased chassis with one of its own Famous-branded engines built at Milwaukee Works.

After examining its options, IH leadership opted to take advantage of existing research and development—one might imagine, somewhat to the chagrin of its resident geniuses, Johnston and Steward—and selected the Ohio Manufacturing Company to supply a chassis and transmission for its new tractor.

The company originally had its tractors built under contract, although IH soon sold more than the Ohio Manufacturing Company's Morton plant could build and began production in two IH factories. Things moved rapidly from that point, with the original design—the Friction-Drive Tractor—refined and improved over the years. The tractors didn't sell in grandiose numbers, but the experience convinced IH that there was a future in tractors.

The first decade of International Harvester tractor production was a confusing time, as development was continuous and chaotic. The company's factories were ill-equipped to build such machines, with space hard to find. In fact, some of the early 45-horsepower machines had to be built in tents due to the lack of space!

The fact that IH was marketing under a wide variety of brands and dealership networks added to the general sense of chaos when one looks at the line of tractors as a whole. Two separate departments were formed to build two independent tractor lines, Titans and Moguls.

In *Century of the Reaper*, C. H. McCormick III states that the company's intention was that the Mogul line would be sold in McCormick dealerships and the Titan line would appear in Deering dealerships. McCormick also wrote that the plan didn't work, insinuating that both lines could be sold in either of the dealerships.

Before the branding shenanigans could begin, however, IH needed to harness the power of its resident machine heads and step into the tractor business.

THE FRICTION-DRIVE TRACTOR

The first production gas tractor offered as an IH was the Friction-Drive Tractor—a product of the prolific pool of early inventors tinkering with gasoline-powered machines. While IH had some of its own people building early experimental tractors at this time—notably Edward A. Johnston, John Steward, and George Ellis—it turned outside the house to build its first production machine.

The chassis was the Morton Tractor, designed by prolific inventor Samuel S. Morton. Morton first built his machine in 1899 in York, Pennsylvania,[40] mating a portable gasoline engine to a cart and making the unit self-powered by a friction drive. He developed the machine to pull a heavy load, and he was granted patents for it in 1902 and 1903. The machine was mainly used to power stationary equipment. Morton moved to Sandusky, Ohio, in 1904 and joined forces with the Ohio Manufacturing Company at Upper Sandusky, Ohio. His patents were transferred to the company in 1906.[41]

A small number of Friction-Drive Tractors were built at Upper Sandusky in 1906. Seyfarth wrote that experimentation began in 1905 and twenty-five were built in 1906,[42] and Baumheckel states that fourteen were shipped.[43] If some of the machines were kept by the factory for in-house testing, they both could both be correct.

Power for the Friction-Drive Tractor was provided by a one-cylinder IH Famous engine. The crankcase was not sealed and used total-loss oiling. The engine was engaged by moving

THE ARKANSAS TRAVELER

This Friction-Drive Tractor was sold in 1906, worked for sixteen years, and then became a showpiece. This photo is dated November 10, 1922. *Wisconsin Historical Society / 68264*

INTERNATIONAL FRICTION-DRIVE TRACTOR

The first IH tractor mounted an IH Famous engine to the friction-drive S. S. Morton, built by the Ohio Manufacturing Company. Counter to the writing on the image, this one was more likely built in 1906. *Wisconsin Historical Society / 4051*

1907 FRICTION DRIVE

This photo shows two men at work with a tractor "built in 1907" and a road grader. The image is dated 1922 and states the tractor was still in use. *Wisconsin Historical Society / 70676*

LONGEVITY

This 1907 Friction-Drive Tractor was photographed in June 1922, and it was still in use by owner Joseph Bamell. *Wisconsin Historical Society / 161191*

it back and forth on rollers. When the engine moved, a 12-inch pulley mounted on the end of the crankshaft was brought into contact with a 50-inch wheel. When the larger wheel turned, it drove the rear wheels through bull gears and pinions.

Keep in mind that the Friction-Drive Tractor weighed in at about 13,500 pounds; one can only imagine the problems of trying to move more than 6 tons of tractor through adverse conditions with friction drive.

Spray or open tank cooling kept the engine cool but lost a lot of water to evaporation. The system uses a pipe with holes, a screened tank, and a catch basin at the bottom of the tank. Water is pumped through the engine's jacket and up into the pipe and out the holes, where it runs down the screen and cools. Most of it is captured and flows back into the engine, where it is pumped through the water jacket and back up the pipe. As you can imagine, water loss due to evaporation was a problem and cold-weather operation a pain.

The Morton frame was constructed of heavy channel iron, with several main beams connected by cross-members. The channel frame was relatively simple to construct but twisted under load. The low horsepower and loose tolerances of early tractors kept a bit of chassis twisting from being much of a problem, but later machines would struggle with channel frames.

The first few Friction-Drive Tractors to roll off the line had tops made of slats with painted canvas and exhaust pipes that ended short of the top. R. W. Henderson, manager of the road engineers with the company, reported that these early machines blew so much exhaust gas, sand, and grit about that they were nearly impossible to operate. The roofs were quickly changed to corrugated tin, with elongated exhaust pipes that extended beyond the roof, which improved operation of the tractor to some degree.[44]

Response to the tractors was good and, for 1907, production of two hundred tractors was authorized. The tractors required improvement—the friction drive was not adequate for plowing and a gear drive was needed. Indicating IH saw potential with the tractors, it transferred its star engineer Edward A. Johnston from Keystone Works to Upper Sandusky in February 1907[45] to develop the gear drive.

Johnston had been working on a three-wheel tractor at Keystone, although that machine was never completed. He continued to develop his own machines after the transfer, building a four-wheel machine powered by a tank-cooled Brennan motor with two forward and two reverse speeds. The Brennan motor was later replaced by a three-cylinder four-cycle motor that was designed for the tractor.[46] According to Earl C. Lutz, general foreman engine assembly, this machine developed by Johnston was completed in summer 1908 and was sent to the contest in Winnipeg. He said after that, the engine design was changed from a 40-horsepower three-cylinder to a 65-horsepower two-cylinder. "The tractor was no doubt the father of the Mogul line."[47]

INTERNATIONAL TYPE A

The friction-drive forward and reverse was found wanting on the early machines, so a gear drive was developed to work better while plowing. This was used on both early Type A and Type B models, which were evolutions of the original friction drive. *Wisconsin Historical Society / 95615*

TYPE A

This variant of the Friction-Drive Tractor was built by the Ohio Manufacturing Company, with Johnston working with them to design a gear drive to replace the friction drive on the original chassis. The design was an exercise in manufacturing techniques, as the specifications from Ohio had such loose tolerances that the gears simply would not mesh.[48] Johnston adapted the design and constructed the Type A gear-drive machine. The tractor used a gear-driven forward and gear-driven reverse.

In 1909, the 12-horsepower, two-speed Type A was introduced. The tractor featured a gear-driven two-speed forward drive and friction-drive reverse, a Famous engine, and stub (independent) rear axles.

INTERNATIONAL TYPE B 20-HORSEPOWER TRACTOR AT WORK

This early IH tractor was used on the farm of J. S. Curtis in Elmira, New York, to plow about 10 acres per day. *Wisconsin Historical Society / 47809*

Two forward speeds had obvious advantages over one, and International describes the tractor as meeting the need for a "fast-moving tractor." Considering the early tractor engines ran at about 240 rpm and propelled the tractors forward at a couple of miles per hour, "fast-moving" was only relative.

TYPE B

The next new variant of the original Friction-Drive Tractor was the Type B tractor, which was similar to the Type A. It used the gear reverse and forward, larger 64-inch rear-drive wheels, and was available only with a 20-horsepower engine. Like the Type A, the Type B was built initially with a single-speed gear forward and reverse; a Type B Two-Speed model with a gear forward and friction-drive reverse was built later.

Both versions of the Type B were produced in low quantities, and only a few are known to exist. According to LeRoy Baumgardner, one of the few collectors who owns one of these rare machines, the Type Bs use a different part numbering system than other International tractors, and IH logos are not found on the existing Type Bs. This indicates that the existing

Type Bs were built at the Ohio Manufacturing Company's Upper Sandusky plant. A few were built in the IH plant in Akron, Ohio, and those would have probably been stamped with IH logos and part numbers.

MOGUL TYPE C

Development of this model began in the experimental department at Akron Works in 1907, as a 20-horsepower machine designed under the eye of Johnston.[49] The tractor was built in 20- and 25-horsepower sizes, and it featured friction-drive reverse, gear-drive forward and single-cylinder engines. The Type C used a single-friction clutch rather than the dual clutches of the A and B two-speeds. The single clutch was spring-loaded to absorb the shock of sudden engagement.

The Type C was the first early IH tractor to sell more than one thousand units in a year, selling 1,787 in 1910. This tractor design was quite different from the Type A and B (although the appearances are similar). The tractor was built at Akron Works for a short time, with production transferred to Milwaukee Works in 1909 or 1910. On January 25, 1911, tractors in this line were renamed the Mogul Type C, making it one of the few early tractors built at Milwaukee Works to bear the Mogul name.[50]

1913 IH TWO-SPEED MODELS
A two-speed transmission was developed for the early IH machines. IH also went away from the gear forward and gear reverse transmission, and it started offering a gear forward but friction reverse configuration. IH advertised friction reverse as smoother to engage with less damage to gears, indicating the gear reverse was problematic. *Wisconsin Historical Society / 23254*

MOGUL TYPE C
The final evolution of the friction-drive platform was the Type C. The Type C was the first IH tractor to carry the "Mogul" brand name. *Wisconsin Historical Society / 49199*

THE WINNIPEG AGRICULTURAL MOTOR COMPETITIONS

Founded in 1891, the Winnipeg Industrial Exhibition drew up to two hundred thousand people to Winnipeg, Canada, to see the latest innovations in farm technology. Two professors from the newly formed Manitoba Agricultural College and an innovative tractor engineer, A. Burness Greig, decided to add a tractor contest to the Exhibition for 1908, both to accurately and technically measure the output of new tractors and also to use the contest as a showcase to convince farmers this new technology could improve their situations.[51]

Greig was an early proponent of the light farm tractor, and his bias was reflected in the first Exhibition, held in 1908—the "Light Agricultural Motor Competition." The tractors had to weigh less than 14,000 pounds, including cooling water and fuel. The weight alone ruled out all but one steam tractor—a J. I. Case 9-horsepower machine.

That first contest took place on July 13–15, 1908, and the tests were for plowing, disking, hauling, and driving a belt. Points were assessed for each machine in another nine categories, including price, range, dirt protection, speeds, and accessibility.

A total of nine tractors entered the first contest—with three of those machines built by IH. They entered two smaller machines—a 15-horsepower machine, a 20-horsepower single-cylinder, and a larger experimental 40-horsepower machine. The latter was an evolved design done by Edward A. Johnston, which he had started working on while at Keystone Works in 1905. The tractor had a friction clutch, two forward speeds, and a chain final drive.[52]

The organizer of the event, Greig, took note of the fact that IH had multiple divisions that competed mightily against each other. "All is fair in love and war, and it is plain to see that each man—though representing the same owners—was waging war to the teeth against the others. It was a competition within a competition."[53]

The 20-horsepower IH machine fared the worst in the competition, finishing sixth out of eight competitors. The experimental 40-horsepower machine did better, finishing fourth, and the 15-horsepower IH won a silver medal for second place. The Kinnard-Haines 30-horsepower machine won the gold medal, edging out IH by just a few points.

In 1909, larger tractors were allowed to enter the Winnipeg contest and the field swelled to twenty-two entrants. IH doubled down and sent six machines, including three that would go head-to-head in the mid-sized tractor class. They hastily put together two Type D prototypes, neither of which fared well in the contest.

The other IH entries, however, performed well. At the end of the day, their 15-horsepower entry was the gold winner in the small tractor category, and IH also won the mid-sized tractor category with one of its 20-horsepower machines.

The Kinnard-Haines won gold in the largest class, edging out a 25-horsepower IH in second, and a J. I. Case won gold in the new-for-1909 steam class (which IH did not enter).

WINNIPEG CONTESTS

One of the prominent contests for early farm tractors was held in Winnipeg from 1908 to 1913. IH entered with enthusiasm—supplying three of the nine machines in 1908. *Rick Mannen Collection*

In 1910, IH again won both small classes, proving it had an edge in small tractor engineering at this time. The new Titan 45 showed itself brilliantly as well, pulling a 10-gang P&O plow with aplomb—the most of any tractor in its class. W. E. Bonner, the owner of the land upon which the contest was held, was so impressed he purchased the unit on the spot.[54]

The following year, IH again had a number of entries. IH employee, A. C. Seyfarth, attended the event and took photographs—he would later write one of the most important texts about the history of early tractors. IH tractors took home two golds for its 20- and 25-horsepower entries. The company took home a silver and a gold medal again in 1912.

After 1912, the Winnipeg contests became less important. In 1913, a major power farming demonstration in Fremont, Nebraska, attracted fifty thousand people. By 1915, tractor demonstrations sprung up all over the country, at county fairs, picnics, and hosted in random fields by local dealers. In 1919, the University of Nebraska began its test program, which remains the gold standard for tractor performance to this day.

1911 KEROSENE IH 25-HORSEPOWER MOGUL JR.

Rain turned the 1911 Winnipeg showgrounds into "Red River gumbo" and made this show particularly challenging. IH entered this early single-cylinder Mogul Jr.

Rick Mannen Collection

EARLY TITANS

Titan tractors were engineered in Milwaukee Works, with several models and sizes built at the location. The first was the Type D 45-horsepower, and development of this machine started as early as 1908. The IH engineers had been experimenting with using an engine developed by other companies, but those proved to be unsatisfactory. IH hired P. R. Hawthorne to design and build a new twin-cylinder engine for the 45-horsepower machine.

Official production was authorized in May 1910, and the tractor was to be built as the 45-horsepower Reliance. It used a heavy channel frame and had the same basic design as the earlier Type A-C models, just on a larger scale. The tractor was to be manufactured at Milwaukee Works in Milwaukee, Wisconsin. Note that in 1911, tractor production was halted at Akron.[55]

The Titan Type D 45-horsepower used a new chassis design that was gear driven throughout. The tractor was significantly larger than previous Internationals, weighing in at 18,500 pounds and standing about 16 feet long and 9 feet wide. The tractor was a radical design for its time. It used removable cylinder sleeves, crankcase compression, automobile steering, and a rear differential.

The tractor's biggest problem was reportedly hard starting. International provided a solution to this problem by equipping the Titan 45 with an air starter. The change took place in 1912 and could not be retrofitted to older machines as it required a host of new parts, including a new set of frame

WORKING MAN'S DETRITUS

The cooling tower on this lovely original condition Titan 45 is covered in graffiti, scratched into the tin by farm workers five to ten decades ago. This one is owned by Lou Buice.

Lee Klancher

1912 TITAN 45

The 45-horsepower tractor first appeared in 1910 and was powered by a twin-cylinder water-cooled engine that used a hit-and-miss ignition and ran on kerosene. The tractor was gear driven (forward and reverse) and featured a bull-gear drive train and rear differential. *Lee Klancher*

TITAN 30-60
The Titan 30-60 was an evolution of the 45 that featured a larger cylinder bore. In 1915, the cooling tower on the 30-60 was abandoned for a more efficient sealed radiator cooling system. Fewer than two hundred of the Titan 30-60s were built from 1915 to 1917. *Wisconsin Historical Society / 7189*

channels and a new front bolster. The number of new parts involved is evidence of how big a problem it was to start early 45s. The air starter used an on-board compressor to store air in a tank. The tank was piped to the right-hand cylinder head and could be released to turn over the engine.

In 1912, a new version of the 45 was introduced, equipped with an improved engine. This machine was similar to the old 45, but with a different crankshaft, flywheel, and a throttling governor.

TITAN 30-60

The next big change came for the 1915 model year, when a new fuel mixer that increased power output was added. The original rating was 27 drawbar and 45 belt horsepower, although it

TITAN ROAD ROLLER
This Titan Road Roller is one of several variants that IHC built. The roller front end bolted onto the stock chassis. The tractors were convertible, and a regular tractor front end could be bolted on for field work. *Henry Hamilton Bennett, Wisconsin Historical Society / 72878*

is more typical to see the original model listed as the Titan 45 rather than the 27-45. The new model brought the ratings up to 30-60, respectively, and the Titan 30-60 was born.

Late in 1914, the 30-60 was redesigned. A fan and radiator cooling system replaced the tank cooling, and a starting engine from the Mogul replaced the compressed-air starting system. A cab was added, some of the controls were moved, and the new machine was fitted with truss rods to reduce vibration. The tractor was released for production early in 1915.

TITAN TYPE D 20- AND 25-HORSEPOWER MODELS

The evolution of the two smaller Titan models began in 1909 at Milwaukee Works.[56] A first rendition of the machine was developed very quickly, in about forty days, built with an eye

EARLY TITAN

Some Titans were marketed with the "Reliance" brand name. The one above is shown in 1916, possibly at the Fremont Tractor Show, a very large exhibition held in Fremont, Nebraska. *Charles Van Schaick, Wisconsin Historical Society / 46216*

1911 TITAN TYPE D

The smaller Titan Type Ds came in 20- and 25-horsepower versions. This machine came from a collector in Washington state, who restored it in the early 1980s then left it outside. Wearing rust and bad paint, the tractor passed through the hands of two other collectors before landing in the stable of Lou Buice, who had it restored by Wendell Kelch. *Lee Klancher*

TITAN 18-35

The Titan 18-35 was put into production in 1912. The engine was a twin-cylinder, cooled by a cooling tower and with an air starting system similar to the 45 and 30-60. Fewer than three hundred of these machines were built between 1912 and 1915. *Wisconsin Historical Society / 161188*

towards the 1909 Winnipeg Agricultural Motor Competition. The machines were painted while en route on the train. "The showings made in the contest," L. B. Sperry wrote of these hastily created models, "did not result in any gold medals for the company."[57]

Thankfully IH had other models at the competition that did quite well. This may not have been satisfactory to Sperry, as the IH engineers competed as much among themselves as with other makers at the show.[58] After the poor showing, the small Titans were redesigned in fall 1909 and ready for testing in winter 1910. The revised Type D Titan 20- and 25-horsepower machines were heavy-duty units, despite their low horsepower ratings. They weighed in at about 14,000 pounds and differed from the Type Cs in their use of a live axle and larger 70-inch rear drive wheels. The tractors were sold as Reliance tractors for a time, and then became known as Titans.

TITAN 18-35

Designed in 1911 by E. D. Eliassen and Mr. Waterman, the Titan 18-35 went into production in 1912.[59] This tractor was a smaller twin-cylinder with two forward speeds and automobile-type steering. Fewer than three hundred examples were built from 1912 to 1914.

The early small International tractors were successful enough to convince IH that tractors were worth developing further, and the machines required significant advances in order to reduce maintenance, improve performance, and become simple and reliable enough for the average farmer to operate successfully. The machines were low powered and slow by modern standards, and most of them used exposed cooling, oiling, and valves that were vulnerable to dust or dirt and required constant maintenance. Visionaries could see that the future of farming lay with powered equipment, but certainly some veteran horse farmers of the day were undoubtedly skeptical of these complex and expensive new machines chuffing and clanking around a rich landowner's estate.

TITAN 15-30

Development of this model began in summer 1912 at Milwaukee Works, with a new four-cylinder engine and chain and sprocket final drive. IH tested the models extensively and kept things as secretive as possible. When the first example was sold in August 1914 to a Mr. Adams, he discovered three coats of different-colored paint on the machine and became convinced he had been sold a used tractor. After a stormy period between Adams and the local dealer, the tractor proved

TITAN 12-25

The new four-cylinder Titan was introduced as the 12-25, with the name changed to 15-30 early in production. *Wisconsin Historical Society / 72998*

1920 INTERNATIONAL 15-30

The machine was known as Titan 15-30 in 1917, and later versions were dubbed the International 15-30. This 15-30 is owned by Lou Buice. *Lee Klancher*

"FLAMING FOUR"

One of the key issues of the day was fuel economy, and the International 15-30 delivered this with water injection. It also started on gasoline and switched to cheaper kerosene once running. To do this, the tractor had no fewer than seven carburetors. The multiple carburetors could spit flame at times, earning the four-cylinder tractors the nickname the "Flaming Four." *Lee Klancher*

its merits and he convinced three neighbors to buy the same model.[60]

These models were known as "light tractors," evidence that IH had begun to realize, or at least consider, the fact that the small tractor was the wave of the future. It weighed a sprightly 9,300 pounds, giving it an unheard-of power-to-weight ratio of 310 pounds per horsepower.

The new four-cylinder model tractor debuted as the Titan 12-25 late in 1914. Shortly after the introduction, the rating was upped to 15-30. Consequently, the Titan 12-25 and 15-30 are one and the same.

One of the key issues of the day was fuel economy, and the International 15-30 delivered this with water injection, and by starting on gasoline and switching to cheaper kerosene once running. To do this, the tractor had no less than seven carburetors. The multiple carburetors could spit flame at times, earning the four-cylinder tractors the nickname "Flaming Four."

In 1916, the tractor engine was redesigned from the original's "L" head to a valve-in-head design. The transmission was also revised.

The 15-30 was plagued with name changes, and the Titan name was dropped in 1917 due to legal action, the result of an antitrust suit filed against the company. In 1917, IH began to rename some products "International." So, the Titan 15-30 became the International 15-30.

In 1921, the Flaming Four was displaced by the much simpler and more effective McCormick-Deering 15-30.

TITANIC ADVENTURES IN SOUTH AMERICA

R. W. Henderson was a road manager for IH, and his job was to go out in the field and help customers with their tractors. This is one of his tales, which took place in Chile in 1912. Note that some Titan tractors were branded "Champion" in South America.[61]

"At Valparaiso, I reported to the office of Savvada Bernard Cia. They furnished me with a German by the name of Hans Brighton to go to a place called Mellapilla to start a 30-60 tractor owned by a man there. I was in Mellapilla[62] until June 1, when I went to Talcahuna to get another tractor that was to be delivered to the same man. We loaded this tractor in a barge to bring it ashore, and as Talcahuna[63] has rather a treacherous harbor, and a norther was blowing pretty severely, the tug crew became excited and cut the lighter adrift. It floundered around in the waves for a few minutes and then sank. The next day I went out in a rowboat and located the tractor which you could see very plainly under about twenty-two feet of water. I started in to make arrangements to have this tractor raised, but found that the insurance company would not let us touch it, so I therefore want back to Mellapilla to finish my work.

On June 11, I left Santiago with a piston and connecting rod for a 25 horsepower Type D tractor located about sixty miles inland from Sliva Scorva. By leaving the train at Mellapilla and going overland I could make this trip in twelve hours, but by taking the train to Sliva Scorva it would take me at least three days. I had been over this road before, so I therefore hired two mules, loaded the piston and connecting rod on one, myself on the other, and started over the mountain pass. The Andes mountains on the Chilean side are exceptionally steep, and on this particular trail, there were places where it was a sheer fall of one thousand feet. In many places the trail was not more than three feet wide hewed out of the solid rock of the mountain. At one of these places, for some unaccountable reason, the mule carrying the piston and connecting rod lost his footing and fell over the precipice. I think he fell possibly five hundred feet before striking the shelving ledge of the mountain and then rolled and dropped at least another three thousand feet.

Through my field glasses I observed that the piston had busted and the connecting rod bent in its fall. The mule, of course, was dead, so I did not undertake to salvage any of the material. It is very seldom that one of these mountain mules lose their footing, but at times they will get frightened at small pieces of gravel and rock rolling across the path, and I presume this is what made my mule fall. When you see a live mule going over and realize what it would mean if you were on his back, it makes you feel rather dubious about taking these trips . . .

On March 27, I was informed that the engine which sank in Talcahuna harbor had been raised and was ready for me. I therefore took a train to Mendoza and hired pack mules to take me across the Andes mountains . . .

Upon arriving in Valparaiso on the night of the 28 of September, I found the entire city excited over a prophesied earthquake.

On October 1, I again went to Mellapilla and stayed with the 45-horsepower tractor there until the 15. This was on account of some red tape from the insurance company, and they would not allow me to touch the engine that was on the docks at Talcahuna . . .

On October 16 . . . I found that my tractor had been setting on the docks for practically six weeks, after being submerged in the water for about the same length of time. The salt brine had even removed every particle of paint from the tractor and had rusted it so that not a working part would move. After tearing the tractor completely apart, I had to use a fifteen-ton screw jack to force the pistons out of the cylinders. I never saw as badly rusted up a mess in my life before. I finally got this tractor running to the satisfaction of the man who had purchased it for about one fourth of its original price, and, on November 1, 1912, I sailed from Valparaiso to Panama.

Case IH

EARLY MOGULS

At the same time as the Titans were being developed at Milwaukee Works, the Moguls were being built, first in Akron and then in Chicago, under the watchful eye of IH's star engineer, Edward A. Johnston.

The Mogul 45 was the first of these, and it was developed at Akron Works in 1908. The Mogul 45 had a radical engine design for the time—an opposed twin-cylinder engine. Production started in a small way at the Blue Island Avenue plant at McCormick Works in 1910.[64] This would become known as Tractor Works on February 10, 1910,[65] and Johnston and his engineering team would come to this location from Akron at the same time.

The facility was small and crowded, so much so that engineers could barely squeeze around the equipment and had to share drafting tables. The tools provided were sparse. "I had no tools to begin with," C. A. Ohlin recalled. "I made up a vise for my own use in the shop out of parts I found in a scrap pile. We surely started under great difficulty."[66]

Tractors were painted in an open-sided shed, large wheels had to be built in an old twine storage room, and one of the large chain hoists was housed in a circus tent. The tent was destroyed in a windstorm on New Year's Day 1911.[67]

Things would improve for Johnston and his engineering team as Tractor Works grew to become a key developmental and manufacturing center for IH tractors. Given the fact the engineers were developing an entirely new type of machine, progress was frantic and difficult. "There were thousands of problems," superintendent C. N. Hostetter wrote. "It has always been a mystery to me how any of the Foreman . . . survived."[68]

About a year after the Titan 45 was authorized, the Mogul 45 was approved. Interestingly, the Mogul's specifications—

1912 MOGUL 45

This original condition Mogul 45 is owned by Lou Buice. Options include the rare lighting unit. *Lee Klancher*

horsepower, size, and weight—were nearly identical to the Titan 45s, yet the design was quite different. It is easy to see that the Titan 45 and Mogul 45 were built by two separate departments, with different ideas about what would work. Some advances crossed over—the starting engine used on the Mogul was later used on the big Titan.

The Mogul used an opposed-twin-cylinder motor with a 9½x12-inch bore and stroke that was slightly larger than the Titans; horsepower, however, was the same, at 45. The Mogul engine revved slightly higher than the Titan's, running at 350 rpm.

Building the new engine caused all kinds of issues at the factory. The size of the castings alone was problematic. The factory was set up to produce smaller parts associated with auto buggies and reapers, so the casting of larger items had to be contracted out. The wheels alone were so large and heavy that a shed had to be outfitted with a large roll, drill press, and a chain hoist. The roof of the shed had to be strengthened to handle the load of hoisting those massive wheels! The spokes and rivets were riveted by hand or with a pneumatic rivet hammer, a time-consuming method for even that early period. The first sheet metal tractor cab was also constructed in this shed.

In 1912, the Mogul 45's bore was increased to 10 inches and the rating was upped to 60 horsepower. The tractor was designated the Mogul 30-60 and was built until 1917.

The big 45s—the Mogul and the Titan—reflected the common belief of the time that tractors would be most useful for heavy plowing and other work requiring more muscle than a few horses could provide. But big tractors were useful only to a fraction of farmers. The machines were expensive to own and operate and, they were effective mainly for plowing large tracts of land and turning a belt all day long. Horses were still much more effective for most tasks, and the big machines were built and sold in very limited numbers.

MOGUL JUNIORS

There were a host of smaller models built during the early and mid-1910s. None of these, including the smaller Moguls and an assortment of Titans, enjoyed the success of later

MOGUL JUNIOR
This Mogul Jr. was shown at the 1915 Paris Exposition. Note the distinctive cooling tower and exposed gear drive.
Case IH

TEST TRACK
Tractor Works had its own test track, shown here circa 1916.
Wisconsin Historical Society / 7639

MOGUL 25-HORSEPOWER

Taken outside of Tractor Works in Chicago, this photo is dated March 25, 1912. *Wisconsin Historical Society / 8561*

small International tractors. These tractors were built in such limited production numbers that they could almost be considered experimentals. The experience gained in designing and building the machines was certainly more valuable than the revenue they generated.

The largest of the small Moguls was the Mogul Junior 25-horsepower. It used a single-cylinder engine that was basically one-half of a Mogul 45 engine. The Mogul Junior weighed 16,300 pounds.

The Mogul Junior 10-20 was also built at Tractor Works in Chicago. A total of eighty-five units were built in 1912 and 1913. It used a single-cylinder engine and was the forerunner to the Mogul 15-30. Both of these models were reasonably successful. The 15-30 was discontinued in 1916, and 527 were built. The Mogul 12-25 was an opposed twin and was produced until 1918, with a total of 1,543 built.

MOGUL 15-30

This is an increased power version of the Mogul 10-20 offered in 1912 and 1913. *Wisconsin Historical Society / 8557*

MOGUL 12-25

This was a new model, designed to be a lightweight addition to the Mogul family, featuring an opposed twin-cylinder engine, two-speed transmission, and chain final drive.

Wisconsin Historical Society / 160526

FUN WITH NEW TECHNOLOGY

R. W. Henderson was sent to demonstrate a new 25-horsepower Mogul Junior at the Springfield State Fair in Geneva, Illinois, in October 1910. The tractor was equipped with the usual canopy and an extra snazzy wooden cab. What followed is best told in Henderson's crisp and entertaining recollection.

"This machine like all tractors was exceptionally hard to start on cold, frosty mornings, and it was necessary to soak rags with gasoline and wrap them around the intake pipe and heat this pipe and mixer thoroughly before we could get the motor started. We had no very good facilities for handling our fuel, so the writer procured two three-gallon candy buckets, one for water and one for gasoline. The gasoline pail had a large red 'G' painted on the side, but this became covered with grease so that it was practically impossible to tell them apart, and one morning Mr. Jarvis filled the water bucket with gasoline. We had just thoroughly warmed the tractor up and got it started, and the rags that were on the mixer were knocked off and fell under the tractor. Being afraid that these burning rags might set the grease afire on the tractor, of which there was quite an extra amount, I picked up what I presumed was the water bucket and dashed the three gallon contents under the tractor to put out these burning rags. Mr. Jarvis was in the cab and had presence of mind enough to throw in the clutch before jumping out. The sight that followed, while expensive in a way, was really beautiful. The tractor walked off across the field dropping a burning board from the cab about every hundred feet. After the cab and wood parts had completely burned off and we had again lassoed 'Barney' as we called her, we spent the rest of the day putting on a canopy and thereafter had to do without a cab."[69]

THE QUEST FOR THE LIGHTWEIGHT TRACTOR

One of the trends that characterized the flurry of tractor development of the mid- to late-1910s was the attempt to build a successful small tractor. Dozens of tractor manufacturers sprung up overnight to attempt to satisfy this perceived need. And most of the agricultural world knew that the company that built a tractor small enough to be used for everyday tasks, while still being powerful and reliable enough to perform them with ease, would make a bundle. The only question was who and when (a simplistic answer to that being Henry Ford in 1917, but that story is reserved for a bit later).

Experimentation was part of the game at IH, and the fact that the company formed through a merger and was managed as a multi-tentacled beast meant a number of sharp-minded gearheads were building interesting machines in basements, circus tents, and shop space carved out of factory floors.

Although International had produced several machines with less than 20 horsepower, all the early models had been nearly as large and heavy as the more powerful machines. Despite this, International had been plugging away at light tractors from day one, with strange little machines crabbing around the yard as early as 1905.

WHITNEY 6-12

The IH records show a Whitney 6-12 tractor, with the lowest production record of any tractor: zero. The Wisconsin Historical Archive has this photo dated circa 1916, and the caption says the tractor was sold in small numbers by International Harvester and by Whitney. *Wisconsin Historical Society / 6968*

MYSTERY MACHINE

This is an interesting prototype. *Wisconsin Historical Society / 7260*

WHITNEY 6-12

This image is also housed in the Wisconsin Historical Archive, showing what is clearly a Whitney 6-12 with IH markings. Other images show the machine driving around. The plot thickens . . . *Wisconsin Historical Society / 161190*

REVERSIBLE MYSTERY

This is another tractor with dubious origins. *Wisconsin Historical Society / 8668*

STEWARD TRACTOR

This elemental creation is another lightweight tractor designed by John F. Steward. Photo dated June 21, 1911. *Wisconsin Historical Society / 22883*

STEWARD TRACTOR

This is a lightweight tractor designed by John F. Steward. Photo dated June 16, 1911. *Wisconsin Historical Society / 22882*

MOGUL 8-16

Introduced in 1914, the 8-16 was a single-cylinder-powered tractor with a distinctive curved frame that rose up over the front wheels, allowing a tight (for the time) 20-foot turning radius. The tractor was quite narrow, at only 56 inches, and smaller than the typical IH tractor of the time. The engine was a horizontal single-cylinder that was hopper-cooled (like the early Titans) and fired by magneto ignition. A planetary gear transmission powered the rear wheels through a single left-side chain final drive. The tractor had one forward and one reverse gear.

The 8-16 sold an unprecedented 5,111 machines in 1915, making up the bulk of IH's tractor production that year. Sales numbers continued to be strong into 1916, with over eight thousand sold. In 1917, the tractor was updated with a larger cylinder bore and a two-speed sliding gear transmission (which replaced the planetary gear transmission) and was dubbed the Mogul 10-20. Sales were strong, but the tractor was discontinued in 1919, probably due to the popularity of the Titan 10-20. This marked the end of the Mogul brand.

MOGUL 8-16

The Mogul 8-16 was the machine that convinced IHC of the viability of the small tractor. Introduced at a time when grain crops were at all-time highs and World War I was driving demand, the 8-16 sold briskly. This one is shown circa 1915. *Wisconsin Historical Society / 73568*

1919 MOGUL 10-20

The Mogul received a bit more power, and the moniker changed to the 10-20. The tractor was quite narrow at only 56 inches and smaller than the typical International tractor of the time. This example was built in the last year of production and is owned by Lou Buice. *Lee Klancher*

SERBIAN MOGULS

The American Red Cross sent thirty tractors to southern Serbia to allow people to open untilled soil during World War I. The tractors are shown here in a photo dated August 12, 1919. *Library of Congress LC-A6195-6395*

TITAN 10-20

The Titan 10-20 was the first great success of the early IH tractor line. The size was right, the timing was right, and—thanks to Henry Ford—the price became very attractive. The tractor was not especially innovative, with the standard channel frame, chain final drive, magneto ignition, and two-forward-speed gear-drive transmission. The equipment was hardly earth-shattering, but it combined all the current technology in a simple, well-built, and relatively reliable package.

Development of the Titan 10-20 began in Fall 1914. By April 1915, the first prototype was plowing fields near Milwaukee, and soon after, two more prototypes were sent to Texas for testing. In the summer of 1915, seven more were built and sold. The first of these created a sensation when it successfully completed a sixty-hour nonstop plowing demonstration at Carlinville, Illinois.[70] The Titan 10-20 was authorized for regular production in 1916, and it was produced at Milwaukee Works.

The Titan 10-20 had a larger-displacement two-cylinder engine, a bit more weight, and more pulling power than the Mogul 8-16. The 10-20 was rated as a three-plow tractor, which was considered the ideal size. Farmers responded favorably to the tractor, and it sold well until 1922, its last year of production. It's fairly amazing that the tractor sold at all in 1922, as the McCormick-Deering 15-30, a tractor that was light years ahead of the Titan, was introduced in 1921.

The Titan 10-20's engine rested in the chassis horizontally, with the two cylinders parallel. Oiling was total loss, meaning that brass and glass oil containers dripped a slow but steady flow of oil onto bearings. Carburetion was by a fuel mixer until 1921, when an Ensign JTW carburetor was put into use. The Titan was started on gasoline then switched over to kerosene, and IH promoted kerosene as the fuel of choice for farming. An IH advertisement from 1919 read, "Kerosene is the practical tractor fuel. Don't let yourself be led away from this fact. Gasoline is an unwarranted extravagance." At the time, gasoline was significantly more expensive than kerosene, a factor that for some users offset gasoline's superior performance.

The twin-cylinder engine was cooled by a thermo-syphon system, which used the engine's heat to circulate coolant

TITAN 10-20

The Titan 10-20 was another step toward the general-purpose tractor. While relatively light, the tractor was larger than the Mogul 10-20, which made it better suited for plowing. Fewer than twenty Titan 10-20s were constructed for field testing in 1915. *Wisconsin Historical Society / 22845*

from the large water tank to the cylinder head jacket. The water tank system was not the most efficient, but it was an improvement over the old hopper-cooled Titans. The engine and transmission were mounted on a steel frame. The tractor weighed in at about 5,700 pounds, which was reasonable at the time.

Power was transmitted to the rear wheels by dual roller chains. Two forward speeds were provided, as well as reverse. The tractors featured small, narrow rear wheel fenders until 1919, when larger full-coverage fenders and an operator's platform became standard.

World War I ended in 1918, but the Titan 10-20 truly went to battle in 1921 (see Tractor Wars sidebar). At that time, the Fordson was dominating the tractor market, and IH slashed prices on the Titan 10-20 to try and compete with the cheap, technologically advanced Fordson. The price for the Titan 10-20 was cut from $1,250 to $1,000 in March of 1921. The price was cut to $900 in October of 1921, and slashed to $700 in February of 1922 when Henry Ford cut Fordson prices to $395, a price that he admitted was less than his cost.

The Titan 10-20 gave IH inklings of the possibilities of the power farming market, and it opened farmers' eyes to the idea that tractor farming could be efficient and economical, even on smaller farms. It was introduced into an era that saw tractor sales growing in almost unbelievable bounds, rising from 21,000 tractors sold in 1915 to more than 200,000 in 1920.

In those fast times, the Titan 10-20 was a staid player, although it performed adequately for thousands of farmers. New technology was on the way, and the Titan 10-20 would go from state-of-the-art to obsolete in less than a decade.

TITAN 10-20 GRADING

In fall 1915, a high-tension magneto was added, and minor improvements were made to the transmission and cylinder design. The Titan 10-20 was released in late February 1916. It was an instant success, with more than two thousand sold that year. *Wisconsin Historical Society / 161184*

HENRY FORD AND THE GREAT WAR

The outbreak of the first World War in June 1914 would show—once again—that war drives increased use of farm technology. The onset of the war overseas drove a great demand for American-made tractors and other farm machines. The European capacity was limited, further hampered by the fact that many of the facilities had trouble securing enough material to build.[71]

With 1915 producing one of the best wheat crops in the first half of the twentieth century, and wheat prices rising 50 percent between 1914 and 1917,[72] farmers were putting as much land as possible into wheat. Worldwide demand for tractors and new harvesting machines exploded.

Meeting the new demand became more complex when the United States entered the war in 1917. IH's German factory in Neuss was seized by the government at that time. The Russian factory simply lost touch with the American arm of IH, which was not able to get any of the profits out of the country. In 1924, the Soviet government took over the Russian IH factory "without payment or promise of one cent of compensation."[73]

HENRY FORD

One of the great innovators of the twentieth century had a terrific interest in tractors. His creations would disrupt the tractor industry. *Public Domain*

WORLD WAR I POSTER

"Wheat will win the war" was a popular slogan during World War I. The massive worldwide hunger for wheat was a catalyst for farm mechanization. *National Archives NWDNS-WW-169B(4)*

During the war, IH helped supply the war effort, building shell-adapters, hand grenades, truck bodies, transport wagons, and artillery wheels. The company also built tractors, harvesters, and other farm machines to help farms supply the world despite a depleted work population.

One of the farm machinery operators who played the war needs to his benefit was Henry Ford, who grew up on a farm and developed a distaste for the drudgery of agricultural life in the nineteenth century. He left school at age sixteen and took a job with Edison Illuminating Company. After a successful career and developing what would be a lifelong friendship with Thomas Edison, Ford founded the Ford Motor Company in 1903. Launching the Model T in 1908 successfully brought to life Ford's dream of affordable internal combustion transportation.

Ford's eye turned to the tractor in 1907, and he built a variety of experimental farm machines over the ensuing

years. The experiments drew a large amount of press coverage and had farmers holding out on purchasing machines until the "new Ford" showed up. Henry's much-ballyhooed entry also had major manufacturers worried about the effects his machine could have on their market. John Deere's caution is particularly well documented.

In 1917, Ford made a deal to supply the British government with a large quantity of farm tractors. He used that purchase to drive his new Fordson, a lightweight and economical tractor that quickly took over the market. The Fordson tractor's first effort was a sale to the British government's Ministry of Munitions, who requested a demonstration. Their initial order was for six thousand Fordson tractors. The machine was introduced to the American market shortly after, with fewer than three hundred sold in 1917. In 1918, more than 29,500 Fordson tractors were sold, accounting for 22 percent of the US market.

When World War I ended in 1918, the war for tractor market was just beginning for IH.

FORDSON TRACTOR

This little machine was light, powerful, and reasonably priced. Farmers loved it—despite its propensity to flip—and the industry was turned inside out. *Library of Congress LC-F82-6650*

INTERNATIONAL 8-16

The International 8-16 was a radical departure for IH, with a sleek, streamlined look, a vertical four-cylinder engine, and a fascinating centrally-mounted radiator. It served as a developmental mule for everything from the power take-off to four-wheel-drive to experiments with rubber tires. The International 8-16 was also the first International to be built on a production line.

The 8-16 four-cylinder model history begins in 1914. The design was drawn at the same time the Mogul 8-16 single was conceived. After early testing, the four-cylinder 8-16 was shelved and the single-cylinder Mogul was produced.[74] Testing of the 8-16 four-cylinder was started again in August 1916.[75] The name was changed to the International 8-16 in 1917.

The transmission was a three-speed unit, and final drive was by chain. A high-tension magneto supplied the spark, and a radiator and fan kept things cool, with coolant circulated by thermosyphon.

1918 INTERNATIONAL 8-16

The International 8-16 was first tested in 1916 as the Mogul 8-16. It was renamed when it went into production in 1917. *Lee Klancher*

A variety of glitches kept it from being produced in quantity until 1918. One of the problems was that the 8-16 used several different engines during its production span, all of them being four-cylinder units that ran at about 1,000 rpm. This resulted in three different serial number series. The first engine had problems with insufficient lubrication. It was replaced, and a new set of serial numbers was assigned. The second engine was apparently underpowered, as it was replaced with a more powerful unit, and a third set of serial numbers was assigned.

The 8-16 was small and light, like the Fordson, and was reasonably priced. It weighed about 600 pounds more than the Fordson, and both used four-cylinder engines with a 4x5-inch bore and stroke that ran at 1,000 rpm. The Fordson was rated as a 10-20 and the International as an 8-16, but actual output and pulling power were comparable, according to tests performed by the Agricultural Engineering Department at Ohio State University. Features like removable cylinders, a multiple-disc dry clutch, and a sliding gear transmission made the International a better-built and longer-lived machine as well.

International 8-16s were originally hand-built. Henry Ford showed the world how well a continuous, moving production line could work with the early Ford cars, and the rest of the industry scrambled to follow. While a mass production line increased capacity and decreased costs, the initial investment required was prodigious.

Part of IH's plan for the new McCormick-Deering tractors was to build them on a production line. The International 8-16 was also built on a production line, at least for a short time. Photos from *Evolution of A Tractor*, published by Stemgas, clearly show 8-16 engines and chain-drive machines being assembled on a production line.

To add an interesting and perhaps confusing twist to the story, early versions of the McCormick-Deering 10-20 were known as the International 8-16 Gear Drive Tractor in company records.

A crude form of a rubber tire was also tested on the International 8-16. This was hardly the first use of rubber tires on a tractor, as the Rubber-Tired Steamer was built in 1871 with rubber-coated steel wheels. The 8-16 used a similar type of tire, as International engineer L. B. Sperry described it in *The Agricultural Tractor 1855–1950*: "The front wheels were fitted with solid-section rubber tires; the rear wheels with

1922 INTERNATIONAL 8-16
The 8-16 used several different engines before a satisfactory one was found. The tractor was not manufactured in volume until 1918, when over three thousand units were built, and it never really reached its sales potential due to late engineering developments that in turn held up manufacturing. *Lee Klancher*

solid-section blocks molded into metal detachable lugs and the blocks were made by the Firestone Tire & Rubber Company."

The tires didn't work well on pavement, and the rubber peeled away from the steel lugs, so the idea never made it past the experimental stage.

The International 8-16 was a hallmark vehicle in many ways. Its full-coverage body work, unique engine and radiator layout, and smooth-running four-cylinder engine were all signs of things to come.

INTERNATIONAL 8-16 DEMONSTRATION
The 8-16 was quite a draw when introduced, but its technical issues slowed sales. Nonetheless, the model was a hotbed for testing, pioneering the power take-off and gear final drive. *Wisconsin Historical Society / 161014*

POWER TAKE-OFF DEVELOPMENT

The International 8-16 was the first IHC tractor to be produced with a power take-off (PTO). The PTO was one of the many innovations of the time tested on the 8-16. It seems that the 8-16 was used as a test mule for all kinds of IHC experiments, including four-wheel drive, six-wheel drive, and crawler tracks. *Case IH*

DEVELOPMENT OF THE POWER TAKE-OFF

By the time the International 8-16 reached the market, the idea of using the tractor engine's power to operate implements had been around for some time. The first clearly recorded example to a power take-off (PTO) was Aveling & Porter's steam reaper, which appeared in the December 19, 1885, issue of *Farm Implement News*. The reaper was pushed in front of the steam tractor, with a chain drive running from the steam engine's flywheel to the reaper. The machine was displayed at the Paris Exposition as early as 1878.

The next reference to a PTO appeared on a tractor built by Frenchman Albert Gougis. Made in 1905, this tractor was a homebuilt machine. Gougis ran a chain from the crankshaft of the engine to a shaft that connected to a McCormick binder. Several universal joints handled flex on the shaft, and the mechanism could be engaged separately from the clutch. In this way, the tractor pulled the binder and powered it. Gougis developed the binder to help farmers save downed grain, and it reportedly worked well for this purpose.

It has been speculated that the idea for the PTO was sparked by Gougis's device, which was spotted during an International executive's trip to France. A photo in the February 12, 1931, issue of *Farm Implement News* shows Gougis operating a McCormick binder with a homebuilt tractor and crude PTO. Ironically, the design of the tractor bears a remarkable resemblance to the Farmall Regular. The photo was taken about 1906, and it can be speculated that International saw more possibilities in the ideas of Gougis than just a PTO.

Engineer Bert R. Benjamin proposed developing PTO drives for the Titan 10-20 as early as 1917. He reported encountering a farmer who owned several binders, a Titan 10-20, a team of horses, and a Happy Farmer tractor powered by an 8-horsepower Cushman engine. Benjamin pointed out that the farmer could afford to purchase several more Titan tractors if they were equipped with a PTO, allowing the farmer to dispense with the expensive and often inefficient powered binders. Benjamin's proposal was received with favor, and a PTO was developed to power the cutter bar and sweep rake lift of the Motor Cultivator. The Motor Cultivator, a specialized machine designed purely for cultivation, failed miserably, but the engineering work opened the door for the Farmall, a tractor capable of cultivation.

In 1919, IH began experimenting with PTO-equipped International 8-16s, and in 1921, the PTO became available as special-order equipment on the International 15-30. The new McCormick-Deering 10-20 and 15-30 also appeared in 1921, and the PTO was available on both machines. The Farmall used a PTO as early as 1922, and it became an integral part of the tractor's design.

Despite the fact that the PTO was available in the early 1920s, it wasn't until later in the decade that it was widely known. The rice crop of 1925, imperiled by a wet season in Arkansas, Louisiana, and Texas, was saved by PTO-driven implements, and the agricultural world woke up to the fact that the PTO had arrived.

EARLY PTO DEVELOPMENT

The power take-off was experimented with after debuting on the IH 8-16. This image shows a McCormick-Deering with what appears to be a live PTO driven from the front of the engine. *Wisconsin Historical Society / 161013*

ELECTRIC POWER TAKE-OFF

An electric PTO was tested on the McCormick-Deering 10-20. *Wisconsin Historical Society / 161011*

ELECTRIC PTO

This is the left side of the 10-20 with the electric PTO. *Wisconsin Historical Society / 161236*

THE ANTITRUST SUIT

When IH was formed in 1902, approximately 85 percent of the harvesting business instantly joined forces under one corporate roof. The end result was a monopoly—this was unquestionably against the law of the time, but not counter to what was happening in American big business. Many other big companies formed that controlled large portions of the market.

This movement in American business took place between 1895 and 1904, with more than 1,800 firms being swallowed up into mergers that created 157 larger companies that controlled at least 40 percent of the market, with forty-two of these new giants controlling more than 70 percent of their market.[76] The bulk of these companies were in a few industries: food, chemicals, petroleum, machinery, and mining.[77] While some of these new firms quickly lost share, a significant number survived to be forces in American industry into modern times. These include names like US Steel, DuPont, International Harvester, American Can, and American Smelting and Refining.

The driving forces for all these consolidations were a complex mix of competition driving companies together, US capital market imperfections making it difficult for small firms to survive, a speculative frenzy in the stock market, and the need for vertical integration in order to maximize profit.[78]

Significantly for IH, the precursor to this movement was Standard Oil, owned by John D. Rockefeller Sr., who happened to be the father-in-law of Harold McCormick. This connection would serve the McCormick family well during the merger, where the access to capital and expertise helped them form the harvester trust and consolidate their family's power within it in the company's formative years. Later on, this connection would prove to be bad publicity, at best.

The American mergers went on at rapid pace around the turn of the twentieth century, some of which were absolutely predatory to consumers and the competition, causing public sentiment to steadily grow against them. The public pressure slowly encouraged the American government to put legislation in place. The states were the first to pass legislation,

TRUST BUSTERS

In 1903, Teddy Roosevelt appointed George Bruce Cortelyou to become Secretary of Commerce and Labor. A darling of the press, Cortelyou served with distinction as Postmaster General of New York, the Secretary of the Treasury, and as President McKinley's personal assistant. He would not, as this image depicts, be able to take down the trusts. *1904 illustration,* Puck *magazine* and IH faced prosecution and repercussions in several states, including Texas and Missouri. The federal government passed the Sherman Antitrust Act in 1890, but it was poorly worded and fell apart when it faced the Supreme Court in 1895.

When Theodore Roosevelt became President in 1901 after McKinley's assassination, he embarked on a complex campaign against trusts. His first suit was filed in 1902 against a large railroad trust formed by James J. Hill, E. H. Harriman, and J. P. Morgan in 1901. Ever the cocky capitalist, Morgan famously told Roosevelt, "If we have done anything wrong, send your man to my man and they can fix it up."

The case moved quickly, but it still took several years to make it to the Supreme Court, where the railroad trust was ordered broken up. Suits were filed against Standard Oil and the American Tobacco Company in the ensuing few years—Roosevelt's administration would prosecute a total of forty antitrust lawsuits.[79]

During Roosevelt's tenure, on Valentine's Day 1903, the Congress formed the Bureau of Corporations (BOC) to study and report on industry—specifically to find and prosecute monopolistic practices. Their reports would be vital to cases against many of the mergers of the day, and the BOC would become part of the Federal Trade Commission (FTC) in 1915.

Roosevelt came to believe that the courts were too slow to be effective, and he favored government legislation over court cases as a means to control monopolies.

Roosevelt also may have carefully selected his opponents. A report created by one of his trust investigators, Burdette D. Townsend, on the harvester trust that recommended it should be prosecuted was suppressed by Roosevelt.[80]

Legislation against trusts accelerated when Roosevelt's protégé, William Howard Taft, was elected in 1908. He ramped up the lawsuits and filed against more than seventy companies, and he doubled down on his efforts as part of a bitterly contested 1912 reelection bid. Roosevelt also ran that year. His presence split the ticket and Taft lost to Woodrow Wilson, who was even more aggressive with antitrust campaigns. Wilson was a personal friend of Cyrus Jr., and while he was highly critical of trusts in his campaign speeches, he never mentioned IH.[81] Despite all the high-powered connections, in 1912 IH finally faced the antitrust suit George Perkins and the founders had feared since they came together in New York in the summer of 1902.

ROCKEFELLER AND McCORMICK FAMILY PHOTOGRAPH

The ties between the Rockefellers and McCormicks were literally genetic, and figuratively just as strong. The two families shared immense wealth and front-page fame tempered with deeply held values of modesty, piety, and frugality. John Rockefeller Sr. (far left) is shown with Nettie Fowler McCormick (far right), Fowler McCormick (front left), and Muriel McCormick (front right). Fowler would become CEO of IH later in life. *Wisconsin Historical Society / 8777*

CYRUS McCORMICK JR.

The son of McCormick Sr. grew up and ran IH with aplomb, seeing the company through the merger and into the turbulent first twenty years of the twentieth century. A friend of Woodrow Wilson, he led the company into great prosperity. *Library of Congress LC-B2-3166-6*

The battle against IH's trust was headline news, with an inquiry done by a house committee. The house committee found the report by Townsend that was buried in 1908, in which he suggested legal action should be taken against IH and pointed out the $5 million payment to J. P. Morgan was "a high price for a simple service."[82] The committee also cited the family connection between the Rockefellers and the McCormicks as well as the evidence of Rockefeller money behind the merger to press the point home that the harvester trust was part of a larger amalgamation of corporate power. Both allegations were oversimplifications but held some truth. And both facts added to the perception of the harvester trust as a deal done by a tight network of rich people getting richer.

The US government's case against IH was filed in St. Paul, Minnesota, on April 30, 1912, and decided against IH on August 12, 1914 (IH immediately appealed to the Supreme Court).

McCORMICK IS OK

The brand chaos at IH continued for decades. This McCormick-centric imagery is from 1910. *Wisconsin Historical Society / 161010*

IH leadership would interpret the positive aspects of the loss in their 1914 Annual Report, stating that the government had found IH was a monopoly—which is why it lost in the lower court—but had not found that IH had abused its monopoly power in pricing to consumers or against its industry competitors.

This was largely true, and the company behavior was fairly benign as mergers go. In fact, in the book *Trusts, Pools, and Corporations*, the harvester trust is listed as a "clean, straightforward combination."

That said, the trial did bring to light a number of unsavory practices by IH in its first decade. Perhaps the most egregious was its purchasing of companies under at-times obvious cover—Osborne and Keystone being the best examples—and then aggressively placing advertising by those companies stressing their market positions as independents.

The company also issued contracts attempting to force dealerships to carry only IH products, bribed state officials, dodged taxes, and received kickbacks from steel manufacturers.

The Supreme Court appeal was argued in April 1915. The justices, however, could not come to an agreement and ordered a rehearing. The case was argued again in 1917, possibly in conjunction with the suit against US Steel, and again no verdict was reached. With the outbreak of war, the suits were postponed due to the need for IH to produce war machines.

In 1918, three IH lawyers worked with the Attorney General to create a "consent decree," which is an out of court settlement in which no judgment is rendered. IH agreed to sell off the Osborne, Champion, and Milwaukee lines of harvesting machines as well as factories at the former Champion plant at Springfield, Ohio, and the former Osborne plant in Auburn, New York. By 1920, the company would adjust its dealer network to have only one per town. IH would drop the appeal to the Supreme Court, and, once the war was over, the government would have the right to reopen the case against IH.

This agreement was made public on August 4, 1918, with the government declaring it a victory. "The government contended that combinations of competitive units on a scale large enough to make the resulting restriction of competition direct and substantial are unlawful without more, on the theory that it was not the policy of the Sherman act to wait until the evils of undue concentration of commercial power have come to pass, but rather to strike at once at undue concentration of commercial power itself."[83]

This principle would become a cornerstone of American law—the mere presence of a monopoly is considered illegal. This has been most recently applied in cases against Apple, Google (Alphabet), Microsoft, and Facebook, as well as determinations that blocked proposed mergers such as AT&T and T-Mobile (2011), US Airways and American Airlines (2012), and Penguin Random House and Simon & Schuster (2022).

The impact on IH was not entirely negative, and some of the conditions didn't even come to pass. While Osborne the company was sold, the buyers didn't want the factory. With permission from the government, IH kept the Osborne plant, as well as the Champion factory. IH renamed each, with Osborne becoming Auburn Works and the Champion factory becoming Springfield Works.

The consolidation of dealerships made sense, and, while the court claims stated there were up to five IH-owned dealerships per town, these were extreme cases and the actual dealerships lost was roughly 25 percent. According to Cyrus McCormick in *The Century of the Reaper*, IH had 21,800 dealers in 1917, and it cut back to 13,860 by 1919.

The more unified branding made sense, and one wonders if IH had done so at day one, it might have been better. The company combined the harvesters, creating the McCormick-Deering line. The new machines combined the best features of both lines. McCormick stated that salesmen were pleased to be selling a better product, that customer service improved, and that the resultant joint experimental department had increased resources and improved focus.[84]

The 1918 consent decree did not end the antitrust suit, and the government would indeed reopen the case in July 1923. The US Attorney General filed a petition to reopen the case in St. Paul, Minnesota, stating the consent decree did not reestablish competition and IH was selling at cost to eliminate competitors. The second clause is rather fascinating—IH was indeed selling at cost, but doing so to match Ford and survive, not to eliminate competitors. This did, in fact, eliminate a lot of weaker tractor makers, but IH was fighting for survival rather than abusing power.

HAROLD McCORMICK
One of Harold's schemes was a Chicago commuter airline, which the windy city quickly doomed with weather that kept flights from being on schedule. The venture lasted only a year. *Library of Congress LC-B2-2712-7*

The St. Paul District Court ruled in favor of IH in May 1925, and the US Government appealed to the Supreme Court, which ruled in favor of IH in 1927.

Ultimately, all these court cases found IH didn't blatantly abuse the power it possessed. Certainly, money was made in great quantities and much of the market was swallowed up by the conglomerate, but the case could not show the ultimate consumer, the farmer, suffered as a result. That was probably one of the most compelling reasons for the government to leave IH nearly intact. The competition suffered, but the consumer did not. And by the time the trial was complete, competition was appearing.

The antitrust suit lasted from 1912 to 1927—a total of fifteen years. Cyrus III would argue it started in 1905, when the state's filings against IH began.[85] The shadow of the antitrust suit lingered at IH for decades. A longtime IH employee who was hired in the 1940s once remarked that the antitrust lawsuit was a topic to be strictly avoided, even into the 1960s and 1970s.

TRACTOR WARS

One of the major challenges faced by International Harvester was the entrance of Henry Ford and his Fordson tractor in 1917. The Fordson was light, reasonably powerful, economical, and sold by a marketing genius through his extensive network of car dealerships. Ford soon owned 75 percent of the market and sold more Fordsons in 1919 alone than most tractor manufacturers had in the past decade.

Ford had very deep corporate pockets and a zealous passion to bring power to the farm. He was willing to cut his margins to nearly nothing in order to sell his tractors.

The tractor industry had to respond to survive, and most manufacturers lowered prices in an attempt to gain some of the market that Ford had created. The low prices drove up industry-wide sales, and tractor use on the farm grew dramatically over the next five years.

While the lowered prices and increased competition brought good things to the farmer, it devastated the industry, and IH was no exception. Almost overnight, IH went from owning the market, controlling 39 percent in 1916, to half of that in 1918.

By 1921, International was fighting for sales and dropped prices several times that year. It also initiated field demonstrations designed to showcase IH tractors and convince buyers that they were worth the extra money. The weapons of choice were the Titan 10-20 and International 8-16, and it seems they were up to the task of taking on the Ford, as IH

FORDSON TRACTOR

When management at IHC realized it could not match Ford's prices and build a quality product, it took the war to the field. Agents were instructed to find out when Fordson sales were about to take place and challenge the Fordson representative to a field trial. The IHC weapons were the Titan 10-20 and the International 8-16. Not the best IHC could build, but the best available at the time. The majority of buyers couldn't refuse a tractor built by Henry Ford for less than $400, and IHC lost precious ground. *Library of Congress LC-F82- 6644*

FORDSON INDUSTRIAL TRACTOR

The Fordson had its share of flaws, but they sold in unheralded volume. Most went to farms, but a fair number were sold for industrial use. *C. J. Hibbard / Minnesota Historical Society SA2.5 p78*

reclaimed a larger share of the sales that year, gaining 12 percent in market share. But Ford wasn't finished.

Ford responded, turning his stranglehold into a death grip by dropping the price of the Fordson to \$395, a price he admitted was below cost. IH cut its prices as well, selling Titan 10-20s for \$700 and International 8-16s for \$670. The company also took the war to the fields, challenging Fordsons to plowing contests.

In 1922, IH was not especially successful, losing market share where a buyer could bring home a Fordson for less than \$400. The rest of the industry fared worse. Seventy tractor companies failed in 1922, and thirteen more dropped out the next year. In a three-year span, the number of tractor manufacturers had been cut in half.

By 1926, IH had its new McCormick-Deering tractors on the market and was regaining the dominance of the past. In 1928, Ford bowed out, and IH's new unit-construction McCormick-Deerings helped it take control of the market; 55 percent of all tractors sold that year bore the IH logo.

International maintained a dominant market position until 1939, when the introduction of a stunning line of new Farmalls was tainted by a handshake between Henry Ford and Harry Ferguson that brought the Ford tractor back to the United States.

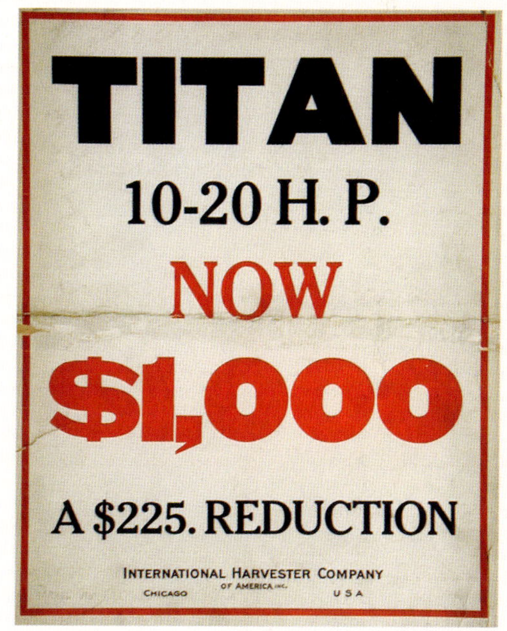

LOW PRICES

In the early 1920s, tractors sales were dropping industry-wide, and Ford and IHC battled for control of the market in what became known as the Tractor Wars. The first part of the conflict was simply a price war. Henry Ford's Fordson was so economical that tractor manufacturers were forced to drop their prices to sell tractors. Tractors built by IHC saw at least four price reductions beginning in 1921 and ending with the International 8-16 at \$670 and the Titan 10-20 going for \$700. Despite this, Ford's 1922 decision to drop the price of the Fordson to \$395 (which was below Ford's cost!) kept the Ford Motor Company at the top of the tractor sales heap. *Wisconsin Historical Society / 96641 (left) / 9277 (right)*

McCORMICK-DEERING 10-20 AND 15-30

The McCormick-Deerings debuted completely enclosed drivetrains, which contributed to their longevity and sucess. They could be serviced easily, due to unit construction that allowed components to be removed easily for repair or replacement. The engine used removable cylinder sleeves, which meant that it could be rebuilt simply, and the final gear drive lasted longer and worked better than the Fordson's worm gear drive. The McCormick-Deerings took the technological innovations that debuted on the Fordson—the integral frame and line production—and assembled them into a high-quality package that was reasonably priced and, with care, would outlive the farmer.

The 15-30 was the first of the McCormick-Deerings to appear. The tractor was a replacement for the International 15-30, with open field and heavier jobs in mind. The McCormick-Deering was a leap to the next level for tractors and had features like ball bearings throughout the engine and transmission, decent carburetion, more reliable ignition, and a lower center of gravity. The power take-off added immensely to the tractor's utility, and it was one of the first tractors that retained its usefulness for a decade or more rather than just a few years.

McCORMICK-DEERING 15-30

The McCormick-Deering 15-30 first appeared in 1923 and was a new machine from the ground up. *Wisconsin Historical Society / 161398*

The two most significant advances on the new 15-30 were the gear final drive and integral frame. International's idea of an integral frame was a large bathtub-shaped frame that housed the transmission and final drive; the engine housing was a separate piece that bolted on. No matter, International's integral frame was many times stronger than the channel frame design, and it was sealed to keep dirt out of the engine and drive assembly.

The integral frame also allowed the use of the 15-30's second key advance—gear final drive. Most previous IH tractors used roller chain final drive. Chain final drive was a less than ideal mechanism, and the gear drive solved the worst of the problems. There was nothing to adjust, the mechanism was fully enclosed and shielded from dirt, and power transmission was positive with little chance for slippage.

The 15-30 was originally called the 12-25 Four-Cylinder International Tractor. The tractor was quickly changed to a 15-30 and geared down a bit. It seems the change in gearing resulted in the power upgrade, as the engine size and equipment were the same. In February 1921, IH ordered that a PTO be made optional on the 15-30. The PTO could be added to tractors already in the field, indicating that some 15-30s were built without a PTO. Vineyard editions of the tractor were built for California in 1922.

Most interestingly, the decision to change the name from International 15-30 to McCormick-Deering 15-30 for the United States and Canada was not made until August 1922. The model was introduced in 1921, making it likely that the first 15-30s were badged with "International" rather than "McCormick-Deering." Only about two hundred were built in 1921, but all of them probably carried "International" badging. Tractors exported off the continent continued to use the "International" name as well as Deering and McCormick badging.

15-30 SPECIALTY MODELS AND ATTACHMENTS

In 1923, rice field equipment was offered for the tractor, which consisted of widened wheels with special lugs and special front-wheel tire ring attachments. McCormick-Deering 15-30s were soon after offered as orchard models, which used lower wheels, a shortened air intake, and had the pulley

McCORMICK-DEERING WITH ORCHARD FENDERS

Designed to be built on a production line, it used an integral frame—meaning the engine, transmission, and rear differential housing were housed in a tub-shaped frame—that was stronger and more dust resistant than previous channel frames. *Wisconsin Historical Society / 161136*

McCORMICK-DEERING WITH ORCHARD BODYWORK

One of the most significant features on the new McCormick-Deerings was the gear final drive, and the original names for the tractors reflected that. The first 15-30s were known as International 15-30 Gear Drive Tractors. *Wisconsin Historical Society / 161135*

McCORMICK-DEERING 15-30

The 15-30 engine had a 4½x6-inch bore, and the crankshaft ran on ball bearings, an innovation that extended bearing life. The engine ran at 1,000 rpm, used splash and gear-pump lubrication, and was cooled by an enclosed system with a radiator. *Lee Klancher*

YARD WORK

This worker is moving a McCormick-Deering No. 8 power-drive harvester-thresher. *Wisconsin Historical Society / 25443*

and pulley carrier removed. The California orchard models, released in April 1924, received the above modifications and rear wheel fenders that covered the top half of the rear wheel and swooped over to meet the hood.

Few tractors were produced in the early years, which may have been a result of early production difficulties. By 1927, production was up to seventeen thousand tractors per year.

In 1926, the McCormick-Deering's transmission was changed to lower the gear ratios for more drawbar pull. The bevel pinion and shaft, drive bevel gear, low-speed gear, medium- and low-speed pinions, medium-speed gear, and reverse pinion and bushing were replaced. The new transmission was not interchangeable with the old.

Beginning in January 1928, rubber-skinned wheels were available for the McCormick-Deering 15-30. Such tires were strips of rubber wrapped around steel wheels, providing some relief to the problem of traversing pavement. Such tires were useful for both industrial and orchard applications. Some orchard growers found that steel lugs damaged roots and smooth wheels provided insufficient traction. These early rubber-coated wheels were less than satisfactory for field work, and until Firestone adapted the pneumatic tire to agricultural use in the early 1930s, lugged steel wheels remained the farmer's most efficient choice.

MORE POWER FOR THE 15-30

In August 1928, the McCormick-Deering 15-30 was given major updates that resulted in a 22-36 power output. The factory continued to call the upgraded model the McCormick-Deering 15-30. Sometime in 1930, the sales department requested the name be changed to 22-36 to reflect the power upgrade. The model was badged as the McCormick-Deering 22-36 for a while, but the name did not stick, and these higher-powered McCormick-Deerings were labeled 15-30s more often than not.

The additional power came from increasing the bore of the engine cylinders from 4½ inches to 4¾ inches. The crankshaft and connecting rods were strengthened, and the cylinder head was new, with larger ports. The cooling system was still enclosed with a radiator, but a water pump was added. The intake and exhaust manifolds were new, and the carburetor was described as using water and fuel, probably some kind of water-injection system. The engine was governed to run 50 rpm higher, at 1,050 rpm. An oil filter was incorporated, and the transmission and rear differential were strengthened. The strengthened parts increased the weight somewhat, and the tractor weighed 6,500 pounds with the standard lugs, pulley, tools, air cleaner, front wheel skid rings, and rear wheel fenders.

The upgraded model and the end of the Fordson in 1928 put some life into sales, pushing unit sales above thirty-five thousand in 1928. In 1933, pneumatic tires became available through special order for the McCormick-Deering 15-30. Production of the tractor tapered off soon after, and the model was discontinued in 1934.

There were some problems with the PTO on the 15-30 and 22-36, as an upgraded PTO shaft and other parts were released in 1928, although the new part was not publicly announced. Some troubles with the rear end also led to several new pieces and improved seals to protect against dirt penetration.

WORKING THE FIELDS

The 15-30 was a larger, more powerful tractor, intended for field and heavy belt work. This one is powering a McCormick-Deering No. 40 harvester-thresher next to an A-4 motor truck on the Foster Farms in Rexford, Kansas, in 1930. *Wisconsin Historical Society / 114665*

McCORMICK-DEERING 10-20

Not long after the decision was made to go ahead with the McCormick-Deering 15-30 (or International 15-30, as it was known in 1921), IH decided to build a smaller model. Like the 15-30, it used an integral frame and gear final drive. The same high-tension magneto, throttle-type governor, gas/kerosene fuel system, combination splash-and-gear pump engine lubrication, and 1,000 rpm engine speed were used. Cooling was the same—an enclosed system with a radiator but no water pump.

Where the new machine differed was simply in size. It used a smaller bore and stroke at 4¼x5 inches. The tractor's dimensions were smaller than the 15-30's in almost every aspect. The 78-inch wheelbase was 7 inches shorter, the 123-inch overall length was 10 inches shorter, and the wheels were 8 inches smaller in the rear and 4 inches smaller in the front.

The 10-20 weighed about 3,700 pounds, compared to the 15-30's 5,000.

While the official decision to build 10-20s was recorded in August 1921, it seems none were actually built until 1923. Note that official or company decisions are documented records of changes and new models. Although company actions may lag behind company decisions, the official decisions are the best source for when new models were built.

Factory records show several related decisions after the initial one from August 1921, all ordering that this new 10-20 be built. The original document is covered in pen marks, as someone apparently wanted the machine a bit smaller than proposed. It also read, "Do not send copy of Decision to Branch Managers until advised by the Sales Department to do so," indicating that the details were not yet satisfactory. The next decision mentioning the 10-20 shows up in

McCORMICK-DEERING 10-20

This tractor is running a belt-driven four-roll corn shredder at the farm of John Bershinski of Chesterton, Indiana, on February 22, 1938. The 10-20 was about thirteen years old at the time, but the owner said, "It runs as well as ever for this belt work—there seems to be no wear-out to it." *Wisconsin Historical Society / 23656*

AIRWHEELS ON A McCORMICK-DEERING TRACTOR
Rubber tires were tested at the Goodyear Tire and Rubber Company on all sorts of machines, including a variety of makes and models of tractors. IH reputedly tested IH 8-16s with rubber; this is a McCormick-Deering fitted with Airwheels at the Cleveland Airport. *Akron University Libraries Archives and Special Collection*

September 1922, over a year later. It specifies that the 10-20, formerly known as the International 10-20, would be badged as the McCormick-Deering 10-20. The report indicates the tractor had not yet been manufactured, supporting the belief that 10-20 production began in 1923. On December 22, 1922, another decision came out authorizing production of the "new" McCormick-Deering machine. Again, the branch managers were not to be notified until the sales department gave the green light. Apparently, they did, because over seven thousand were produced in 1923, the 10-20's inaugural year.

In 1926, the Narrow Tread McCormick-Deering 10-20 was approved for production. Intended for use in orchards and other places that required tractors with a narrower tread, it had a 50-inch rear tread width, which was about 10 inches narrower than that of the standard 10-20. Altering tread width required new rear cases and significant costs for new tooling, changing production, promotion, and so on. Because of this complexity, IH offered only two different tread width 10-20s to suit farmers' needs. This was a problem IH would look to solve with the Farmall.

In 1933, the pneumatic tire became available on the McCormick-Deering 10-20 through special order.

INDUSTRIAL MODELS

About the time the new McCormick-Deering machines were being introduced, IH began to market and sell industrial versions. Part of the company's effort was to get articles into newspapers and magazines touting the money industrial tractors saved cities and townships. The articles were written with a slant that is somewhat entertaining today; IH didn't pull any punches with press coverage. This example comes from the *Chicago Daily Tribune* of June 25, 1924:

"It has been demonstrated in dollars and sense that skinny old nags and bony old plugs, some lame and some blind, are too expensive to carry on the city pay rolls, even if they have aided many a captain to carry his precinct in a primary or an election."

The tactics worked well, and industrial tractors became a long-standing success for IH.

MODEL 20

The McCormick-Deering 10-20 was adapted for industrial use in 1922. The Model 20 serial numbers are indistinguishable from other McCormick-Deering 10-20s. Cast iron disc wheels were used front and rear, and only twenty-three of these machines were built in 1923. A foot accelerator, suspended front end, high-back seat, underslung muffler, dual rear tires, and rear-wheel brakes were added in 1924, with only modest sales increases.

Development continued for 1925, and a high-speed (10.4 miles per hour) top gear and transverse leaf front suspension were added. Front and rear bumpers and an assortment of wheel lugs were added as options. Sales for 1925 are listed as "several hundred" and IH stated, "there was considerable market for this tractor in modified form."

The Model 20 used a four-cylinder valve-in-head IH engine. Top speed was about 10 miles per hour, with the three lower gears pulling 2.3, 4, and 7 miles per hour. The intake and muffler were mounted to the left side and didn't protrude much higher than the hood, presumably for ease of use on the factory floor.

The Model 20 continued to be built through 1940, and the first Nebraska Tractor Test of the machine took place in July 1931. It put out 23.01 horsepower at the drawbar in low gear and 29.87 horsepower at the belt pulley. As tested, the tractor weighed 5,415 pounds and was equipped with an IH air filter, governor, and E4A magneto. A Zenith C5FE carburetor was used.

CITY OF NEW YORK INDUSTRIAL

The 10-20 Industrial was the beginning of a successful new line for International. Agricultural tractors could be converted for industrial use with little more than a change of wheels, and the tractors proved immensely useful for a wide variety of industrial applications. *Wisconsin Historical Society / 161138*

McTRUCK

This interesting industrial experimental fit a truck cab to a McCormick-Deering tractor. It was never produced, which is probably just as well. *Wisconsin Historical Society / 161137*

MODEL 30

This is the Model 30 Industrial, which was simply a McCormick-Deering 15-30 with industrial wheels, shown working in September 1927. *Wisconsin Historical Society / 131381*

MODEL 30

A prototype industrial version of McCormick-Deering 15-30 was constructed early in 1929. The "new" machine was dubbed the Model 30 and had the same general features as the Model 20. Production began in 1930, and very few of the machines were built.

An IH industrial tractor brochure circa 1930 lists the Model 20, Model 30, and I-30. Looking closely at the specs, it appears that the Model 30 is indeed little more than a McCormick-Deering 15-30 with some special equipment. Company records indicate that the Model 30 serial numbers were simply 15-30 numbers stamped with the prefix "HD."

The Model 30 used a higher-powered version of the 4¾x6-inch four-cylinder engine used in later McCormick-Deering 15-30s. The Model 30 engine was rated for 45 horsepower and was governed to run at 1,050 rpm. Production ended in 1932, with only 532 Model 30s constructed. This extremely low number makes the Model 30 one of the rarest IH tractors built and a great find for the collector.

AUSTIN WESTERN GRADER

This photo labeled June 16, 1926, shows a state highway commission worker operating an Austin Western grader built around a McCormick-Deering tractor. The worker is leveling road #2 east of Sigourney in Keokuk County. *Wisconsin Historical Society / 7259*

EARLY IH CRAWLERS

Tracks had been used on production machines in the US since Holt's crawler in 1906, and the traction advantages were clear to manufacturers. IH experimented with tracks on several early machines, including the International 8-16 and motor cultivators. It also adapted French & Hecht aftermarket tracks to the McCormick-Deering 10-20 and 15-30 in the 1920s.

IH began adapting the 10-20 to tracks in the mid-1920s.[86] While experiments were started on the smaller 10-20s, a few larger McCormick-Deering 15-30s were fitted with tracks, but the larger machine was eventually dropped. The smaller model was completed early in 1930 and was put into production for 1931 as the TracTracTor T-15, with fifty machines built. In 1932, the model designation was changed to the T-20.[87]

The TracTracTor T-20s were built at Tractor Works. Problems with the early TracTracTors centered on track roller and bearing wear, mainly because they ran down in the dirt. Steel guards were built to cover the lower rollers, but the ultimate solution would be sealed bearings.

The TracTracTor T-20 was tested at Nebraska in October 1931. It weighed 7,010 pounds and was rated at 18.33 drawbar horsepower and 26.59 belt horsepower. The carburetor was a Zenith K5.

The crawler equipment made the TracTracTor about 2,000 pounds heavier than the McCormick-Deering 10-20. The additional weight and the traction afforded by the treads let the TracTracTor pull about twice as hard at the drawbar as the McCormick-Deering. The TracTracTor compared favorably to its crawler competition as well. It was about 1,300 pounds heavier than the comparable Cletrac, the Model 15. The IH machine outpulled the Cletrac by about 20 percent. The TracTracTor also outpulled the Caterpillar Model 20 at the drawbar, although the Cat pulled at a slightly higher speed. Despite the TracTracTor's respectable showing, it was never produced in quantity. The machine convinces IH that crawlers were a viable product, and the company continued to develop new models.

MOON TRACKS
International began experimenting with tracked versions of the McCormick-Deerings in the 1920s. This circa-1924 machine is equipped with tracks built by the Moon Track Company of San Diego, California. *Wisconsin Historical Society / 24202*

TRACKED 10-20

This Moon Tracks-equipped McCormick-Deering 10-20 also has electric lighting. *Wisconsin Historical Society / 161012*

TRACKED 15-30 EXPERIMENTAL

A very small number of 15-30s were built with tracks. This one weighed in at 10,220 pounds, and this photograph is dated February 21, 1930. *Wisconsin Historical Society / 129298*

IH EARLY CRAWLER

The first crawler model produced by IH was the No. 20 TracTracTor, shown here in a photograph dated June 2, 1930. The weight was listed as 7,430 pounds. *Wisconsin Historical Society / 12129*

H. O. TOLEDO, O.

THE MECHANIC
Wisconsin Historical Society / 133822

CHAPTER THREE

THE FIRST FARMALL

1923–1932

"While going among tractors on the show grounds, I was impressed with the fact that ours are more crude than any."
—John F. Steward, lamenting the sophistication of IH tractors in a letter to C. S. Funk, September 9, 1910

Designers and promoters and all sorts of people flooded us with orders to build tractors for them. They wanted big ones—the biggest possible. The bigger they were the more money they could get for them."

While this refrain could be from the 1960s, the 1980s, or, for that matter, last week, this grievance was written in 1916 by colorful ag equipment columnist Barton Currie. He went on, offering that the large machines popular in the teens were expensive, unreliable, started poorly, and were a fiscal disaster.

"It required the services of an entire community to start one . . . Repair bills on a single tractor would range from $500 to $1,500 in a season. Farmers went broke, manufacturers went broke, and distributors went broke."[1]

The early tractors reviews were not always favorable, and the gloom and doom was reflected by a 1910s survey that found "nine out of ten bankers predicted the [tractor] was the first step to bankruptcy for the farmer."[2]

At the time of Currie's article, demand for the big tractor was taking a pause, but the interest in smaller machines was exploding. In fact, there was a mind-boggling rush of new technology flooding the farmer with leaflets, flyers, advertisements, and door-to-door sales calls. More than 170 new companies had

THE IVEL TRACTOR

Engineer and cycling racer Daniel Albone designed, built, and sold bicycles, cars, and a farm tractor that he debuted in 1902. The tricycle design was revolutionary at the time—and a predecessor to the Farmall. The various Ivel machines, all using the tricycle design, won several motor trials in the UK and garnered significant press attention. Albone died suddenly in 1906, possibly of a stroke, at only forty-six years old. A licensed version of the machine appeared in the US as the unsuccessful Hart-Parr Little Red Devil. Without Albone, the company lost its way, and, by 1919, it was gone. *Science Museum Group*

MOLINE UNIVERSAL

The Universal Tractor was developed by an Ohio farmer, who had running examples in 1913 and 1914. The tractor's design was unique and revolutionary, and the machine was the first practical example of a general-purpose tractor. IH engineers later referred to the Universal Tractor as an inspiration to their work, and the company considered purchasing the machine for resale. IH passed, the tractor was purchased by Moline Plow Company in 1915, and the tractor had a run of good sales. *Wisconsin Historical Society / 92860*

1915 INTERNATIONAL EXPERIMENTAL GP TRACTOR

According to Edward A. Johnston, this is the first attempt by IH to build a general-purpose tractor. The experimental machine was designed by Johnston and IH engineer David B. Baker and demonstrated for Alex Legge and others in fall 1915. *Wisconsin Historical Society / 24651*

DOWN ON THE FARM

This sheet music cover is for a song that demonstrates that soldiers returning to their farms after World War I were ready for new technology and more prosperous times. The new tractors were available in droves—more than 170 makers offered them for a time—but the good times for the farm economy were fleeting. A short period of strong prices and demand would start in 1920 and then fall flat. *Creative Commons / Albert Wilfred Barbelle*

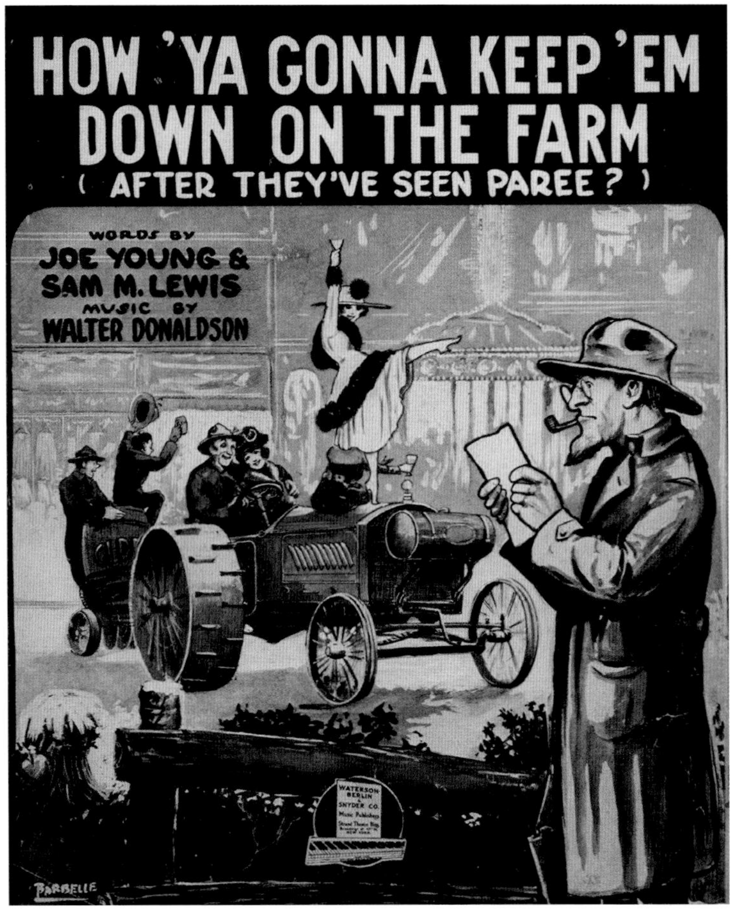

popped up to build small tractors. The race was on as the world realized that the future of farm technology was the machine that could replace the horse. "The great and almost overwhelming demand for small tractors—a cash demand, mind you—is blazing the way as nothing else did."

Currie wrote that line during the boom of 1916—the welder-wielding visionaries of the world knew long before the way things would unfold, and they had been cutting and grinding to that effect for years.

IH's resident engineering wunderkind, Edward A. Johnston, had been hard at work designing tractors (or things much akin to them) since 1900. The Mogul line was developed under his watch, not to mention with his patents.

He had started his agricultural technology journey with small machines—notably the Auto-Mower. From 1912 to 1914, he and his engineering team at Tractor Works were lasering in on creating the tractor that could replace the horse, with "much discussion and many sketches covering a general-purpose tractor."[3]

Other early leading lights exploring this avenue include H. C. Waite, who developed the Waite lightweight tractor in 1913. The idea was sound but the execution flawed. IH sold very few of these, and the tractor was "not successful."

ALEXANDER LEGGE AND THE MOTOR CULTIVATOR

Although the Motor Cultivator was a failure on the sales floor, it was a critical test bed in the development of the general-purpose tractor. Alexander Legge (center) and Harold McCormick (right) are shown with an experimental model in 1919. *From the book,* Alexander Legge

The Universal Tractor was developed by a farmer in Ohio, also in 1913. This model was offered to IH as well, which turned down the opportunity. The Moline Plow Company picked it up and sold it. The Moline Universal would see a strong surge of demand in 1919.

In 1915, the first IH general purpose tractor—the precursor to the Farmall, if you will—was developed by Johnston with help from David B. Baker and others from Tractor Works. The crude machine was demonstrated to Alexander Legge and other IH executives in the fall of 1915.

Four years later, according to Johnston, the first Farmall that resembled the production design was assembled by Johnston and Baker. The first drawing of this machine was made November 10, 1919, with prototypes constructed.

In 1916, Bert R. Benjamin began designing and overseeing the designs of implements for the Farmall. It wasn't until 1920 that Benjamin—the widely-credited "father of the Farmall"—would propose that machine become a replacement for the horse.

This is not to take anything away from Benjamin—it is just a reminder that inventions do not emerge spontaneously. They are evolutions, and the Farmall design began long before Mr. Benjamin took the design under his wing.

In fact, Johnston credits someone else with the original concept. The vision for the Farmall, it appears, came to International Harvester when one of the crankiest men in company history visited a large display of tractors at a 1910 state fair.

His critical eye (and acerbic writing) disparaged the IH line in a widely-circulated letter. In addition to his criticism, the cantankerous genius zeroed in on the potential of a rival maker's design to build the revolutionary machine the entire industry was seeking: a small tractor that could truly replace the horse.

THE FOUR-ROW FARMALL
One of the more efficient and popular implements designed for the Farmall was the four-row cultivator. Benjamin developed this idea after examining Farmall QC501—the first 1924 production model—at Taft, Texas, in 1924. The four-row cultivators were first demonstrated at Hinsdale in June 1925 and featured automatic shifting of the cultivators as well as automatic depth regulation. *Wisconsin Historical Society / 131369*

THE FIERCEST OLD-TIMER AND HIS LOST CONNECTION TO THE FARMALL

By Guy Fay

One of the key players in the Farmall story started his tractor engineering career long before the Tractor Works people started running around odd-looking machines in Chicago. John F. Steward was a legend by the time he started experimenting with tractors in the early 1890s.[4] A man of many interests, he served in the Civil War and participated in the first advances beyond the grain reaper. In a Deering Company that had its conflicts and characters, he was pretty much the most cantankerous character of all. On a rafting expedition down the Colorado River through the Grand Canyon, his war injuries had him in pain; he became very sick and had to be carried out on a pack train. He managed to survive, possibly by being too mean to die.

After he returned, he worked with George Ellis, who had been hired by Deering to devise a way of using domestic fibers to replace the expensive foreign-produced fibers they relied on at the time to bind grain. Together, Ellis and Steward started experimenting with gas engines circa 1889. From the start, their experiments did not focus on the large steam traction engines of the era; they mated small engines to farm equipment to make lithe, agile, integrated farm machines as early as 1891.[5] While none of these creations entered production, they received plenty of publicity and were widely copied by other inventors. The

JOHN F. STEWARD
One of the most inventive minds to come into the IH engineering ranks was Steward, an engineer who came from the Deering Harvester Company. Shown here circa 1900, he would head the patent department for IH, despite a long history of sharp criticism of the McCormick family. *Wisconsin Historical Society / 92226*

HACKNEY AUTO PLOW
Steward spent several hours examining a Hackney tractor at the 1910 Minnesota State Fair. His observations, documented in a long and somewhat cranky letter, would help IH find its way to the Farmall. *Lee Klancher Collection*

groundbreaking Deering Automowers, to some degree developed in the Steward family basement, made history.

Steward took a trip to the Minnesota State Fair in 1910, a journey that became a pivotal and largely ignored moment in IH history.[6] After spending a considerable amount of time at the famous Machinery Hill display, Steward wrote a scathing letter to several other members of IH management, lamenting IH's emphasis on heavy, primitively-constructed machinery and unfavorably comparing IH's efforts to select other manufacturers.

He made three visits to the Hackney tractor, spending two hours with the tractor inventor and another representative. The machine featured a four-cylinder automotive-style engine on a chassis with the drive wheels in front. The tractor worked, but Steward stated that the inventor was still "at sea."[7] Steward noted that the owner of the company, Leslie Hackney, had not invented the tractor but held the patents to the machine. He also noted that Hackney possessed the finances to build, produce, and sell it.

The Hackney's layout and the mounting of the implements were of greatest interest to Steward. The tractor was triangular, with the two drive wheels in the front and the steering ground wheel in the back. The implements were specific to the tractor and mounted underneath, where their weight contributed to traction and allowed them to be more visible to the operator. Steward was highly enthusiastic about this and stated to IH management that this was the tractor of the future.

While the patents for the implements were assigned to Leslie Hackney, the inventor was John Froelich. After Froelich and Waterloo Gasoline Engine Company parted ways, Froelich had gone to the Iowa Gasoline Fire Engine Company in Dubuque, where gasoline fire pumps and other products were made. Froelich stated to newspapers that a traction engine was underway at this company.[8] He also went to work for a foundry in Marshalltown, Iowa, in 1900. While working there, he said the foundry was considering production of a new automobile with a very quiet engine that he had just designed.

Meanwhile, Froelich's brother had gotten into washing machines. Farm wives were desperately in need of labor-saving devices and Froelich Manufacturing was established in St. Paul. John Froelich moved there, possibly in 1902; he went through bankruptcy in 1905.

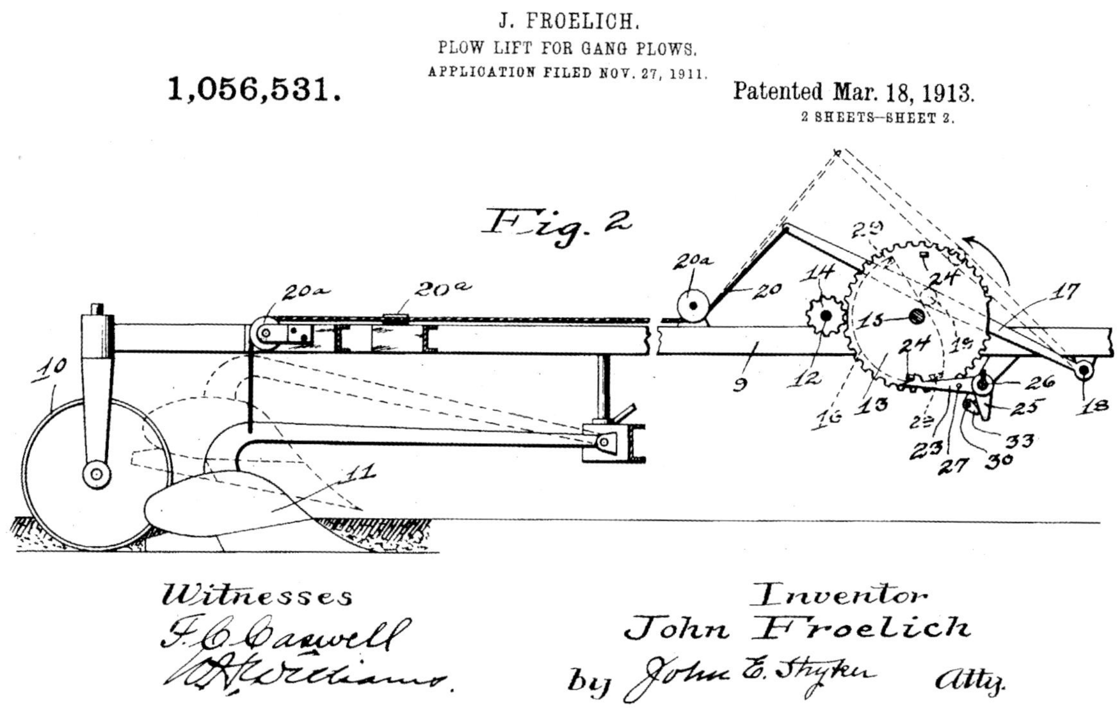

HACKNEY PATENT
John Froelich—better known as the inventor of the Froelich tractor—filed this patent for the plow lift on the Hackney Auto Plow. J. F. Steward spotted the innovation at the Minnesota State Fair in 1910. *US Patent Office*

John Froelich was the co-inventor of a dishwashing machine and started a company to manufacture it in 1906. The dishwasher did not go far, but his brother's company was displaying washing machines on Machinery Hill at the Minnesota State Fair for years afterwards.

At some point, Froelich did some side work for Leslie Hackney, designing several implements that carried his name on patents. The implements, including a push binder, were true advances in the field. This was not lost on the extremely experienced John F. Steward, who returned to International Harvester and started singing the praises of Hackney's implements and doing some experimenting of his own.[9]

Steward's Hackney (aka Froelich-inspired) tractors were soon out in the field working with mounted implements. The location of the implements and the wheel configuration were credited in internal IH tractor histories written twenty years later as the inspiration for the development of the Farmall.[10]

Steward would pass away in 1915. Froelich would later go into the savings and loan business and die in the 1930s in obscurity. He may not have ever known about his impact on the creation of the Farmall. There is no evidence to indicate that he talked to Steward at the fair in 1910, as he was probably stationed at the Froelich washing machine display.

IH AND THE McCORMICKS ROAR INTO THE 1920s

As IH engineers struggled to find the innovations and resources required to solve the puzzle of creating a general-purpose tractor, the company and, for that matter, the country, spiraled through a decade that was easily as tumultuous as the development cycle of the small tractor.

Cyrus McCormick Jr. had successfully led the company through the difficult decade of backroom bickering and consolidation after the merger. While James Deering represented his family with a position on the board of directors, the Deering family was not part of the company officers who ran the day-to-day operations of the world's largest agricultural machine producer.

With marital, personal, and fiscal ties to the world's richest family, the McCormicks had taken over the lion's share of control of IH and made themselves fabulously wealthy as well.

When *Forbes* magazine released its first annual list of the wealthiest Americans in 1918, Cyrus McCormick Jr.'s $60 million net worth ($1.2 billion in 2023) placed him 18th on the list. His ally—Harold McCormick's father-in-law—John D. Rockefeller, topped the list.

By 1919, Cyrus Jr. was ready to slow down a bit. Harold's airline venture from 1914 failed, and he stepped in to help his older brother by serving as IH president, a duty he somewhat ruefully accepted and would publicly admit was not "fun."[11]

CYRUS JR. AND SONS
By 1918, Cyrus McCormick Jr. (center) was one of the twenty richest people in America (at least according to *Forbes* magazine), with his family's personal lives generating front-page coverage nationwide. He is shown with his sons, Gordon (left) and Cyrus III (right), circa 1915. *Wisconsin Historical Society / 77376*

THE BOOM
By the time this photo was taken on August 31, 1928, the problem with the Farmall was meeting demand. *Wisconsin Historical Society / 121023*

The ensuing decade was one that would see IH turn its market domination into dollars, despite a farm economy that was more volatile than usual. The year 1920 would prove a good example, kicking off with demand that outstripped the IH supply of machinery, following with tanking prices that turned the boom to bust. The result was record-breaking sales at IH that led into 1921, the worst year in the history of the agricultural business to date.

In December 1921, Harold's marriage to Edith Rockefeller imploded publicly. In 1922, Harold married Ganna Walska, a thirty-five-year-old Polish opera singer. He also had an operation with Dr. Victor Lespinasse, a sophisticated and high-profile Chicago urologist huckster who transplanted glands from primates into people to increase virility, fertility, and more.[12] All three of these events were widely covered in the media—the gland transplant was emblazoned on the front page of the *New York Times* on June 19, 1922. The *Times* piece went so far as to speculate that Harold had paid a premium and had glands from a human donor implanted.

A few weeks prior to this, Harold had stepped down as IH president[13] and would spend most of the ensuing decade assigned with the rather nebulous title (and role) of "Chairman of the Executive Council."

He was also removed as an officer of the company—the group of people that managed day-to-day affairs. Harold's demotion meant Cyrus Jr. was again an officer of the company, serving as vice president. Harold's implosion also gave one of IH's long-standing stalwarts yet another moment to shine.

When Harold was ousted (or quit), he named his successor as none other than the former Nebraska cowhand, Alexander Legge. His dependable hand had steadied IH since it was formed, and his influence reached deep into the organization. His intelligent leadership would play a key role in the survival of the Farmall, which would face a difficult road in the coming years.

In the early 1920s, IH would improve the efficiency of its manufacturing facilities and begin to face growing pressure from labor. Its systems favored the most productive employees with the best pay, a practice that would be widely decried as inhumane and would later be phased out as the labor unions rose in power. The systems were not good for people, and often unfair, but they produced tremendous profits.

THE COWBOY STEPS UP

Alexander Legge, shown with a pitchfork in his hand, would become company president in 1922. A western Nebraska farm kid without a college education but sharp as a tack, Legge was one of the key drivers of innovation at IH. *Wisconsin Historical Society / 11477*

In the mid-1920s, the farm economy was on the rise and prices would steadily increase, as would IH profits and sales. Foreign sales would hit a record high in 1926, and gross sales and profits set company records in 1928 and 1929.

As the decade closed, Alexander Legge would be called on to serve on the Federal Farm Board on July 1, 1929. The board of directors was filled with names of those who had been with the company since it formed. Most of their terms were due to expire in the early 1930s.

As 1930 loomed, the IH guard and the times were changing. But before that change could come, the company would develop the single most important farm tractor in its history.

THE MOTOR CULTIVATOR

One of the oddest and least successful machines in agricultural history—the Motor Cultivator—would also prove to be a key step in the journey to the general-purpose tractor. The idea for the machine was simple; typical tractors of the day could plow and turn a belt but were not able to cultivate. The Motor Cultivator was developed to meet this need and do one thing well: it was a self-propelled machine that would cultivate crops.

The Avery Company was an early tractor manufacturer known for innovation, and it billed itself as the "world's largest tractor company," employing 2,600 people. Avery offered the Louisville Motor Plow, which was a motor grafted on to one of its cultivators. They produced it from 1914 to 1917. The Universal, developed in Ohio and sold to Moline, was a similar if slightly more sophisticated design that appeared in 1913.

The first IH motor cultivator was designed at Tractor Works by Edward A. Johnston, Carl Mott, David B. Baker, Philo Danly, and others. IH demonstrated its first machine in 1915 and applied for patents on a design for a Motor Cultivator in 1916.[14]

The International Motor Cultivator evolved significantly over time. The basic configuration was a four-wheel tricycle design, with the two rear drive wheels close together and mounted on a swivel. The machine was turned by pivoting the rear wheels. Out front were two tall, narrow wheels, spaced widely apart. The cultivator hung below the chassis and between the front wheels. The designers experimented with lots of variations, including tracks and a wide variety of attachments.

Early tests of the Motor Cultivator showed that it would overturn when driven on side hills. A nose weight partially solved this problem, but the machine still tipped over too easily.

The Motor Cultivator fared poorly in the fields. It was slow and underpowered, and the wheels broke corn roots. Under ideal conditions, the machine performed adequately. Ideal conditions were relatively rare on the farm. The Motor Cultivator's questionable performance and short supply effectively

EARLY MOTOR CULTIVATOR
According to Edward A. Johnston's recollections, the first Motor Cultivator was built at Tractor Works by Johnston, D. B. Baker, Carl Mott, Philo Danly, and others. This experimental is shown circa 1914. *Wisconsin Historical Society / 46214*

1918 MOTOR CULTIVATOR

The Motor Cultivator production was very limited. The machine was powered by a four-cylinder engine that was water cooled with a radiator and fan that used thermosyphon circulation. The engine was mounted to a channel steel frame, and the entire unit weighed about 2,200 pounds. Improvements were made to production versions to try make them less tippy—which largely didn't work. *Wisconsin Historical Society / 161388*

TRACK-LAYER MOTOR CULTIVATOR

Experimentation with motor cultivators continued into late 1920s, eventually evolving into the Farmall. Later experimentals had both a PTO and a lift. This tracked model is fitted with a corn harvesting system. *Wisconsin Historical Society / 161180*

1920 EXPERIMENTAL CULTIVATOR TRACTOR

This evolution of the motor cultivator features a rigid rear frame topped with a transverse-mounted engine that drove the wheels with right and left chain drives. This experimental model is shown in December 1920, testing a cultivator that proved too awkward to develop further. *Wisconsin Historical Society / 22763*

killed it for 1917. By the time the factory assembled about one hundred machines, it was so late in the season that the sales staff felt it could not sell the remaining stock of just over fifty machines.

In 1918, heavier cast rear wheels were added to the Motor Cultivator. The rear wheels were added to offset the machine's tendency to tip. The wheels helped, but the design was inherently unstable. The engine was carried high on the chassis. If the Motor Cultivator was turned downhill, it would tip over.

To worsen matters, the 1918 model was delivered late. Dealers were promised 150 models by June 10. They didn't show up until July 15.

In August 1918, the final nail was placed in the Motor Cultivator's coffin. Cost estimates for the 1919 model were high, at over $500 for a machine that would retail for about $650. Revised cost figures came in, and the retail price would have had to be over $800 for the company to make a profit. Avery's motor cultivator sold for $540; $800 was too high for the market to bear.

In 1919, the decision was made to sell off what was built and discontinue the Motor Cultivator. Edward A. Johnston and others saw the potential of the design, however, and continued development.[15]

The development done on the machine—particularly the interesting developments later in its life, which included power lifts, power take-offs, and implements directly connected to its belly—would prove invaluable to IH engineers going forward.

The failure of the machine, however, would create challenges. In the report "Notes on the Development of the Farmall Tractor," C. W. Gray wrote: "There is good reason to believe that the abortive attempt to market the Motor Cultivator all but completely discouraged the management in its purpose to build an all-purpose tractor."

THE FATHER OF THE FARMALL

Bert R. Benjamin grew up on a farm. Like so many farm kids, he started helping out at a young age—he recalls being eight years old when first put to work.

By the time Benjamin was twelve years old, the young Iowa farm boy could adjust the shares on a two-furrow sulky plow, set the tines on a spring-tooth harrow, and sharpen the knife on a mower. Born on December 17, 1870, in Newton, Iowa,[16] the mechanically gifted lad soon became known throughout the district as one who could fix any broken farm machine.[17]

By the time he was fourteen years old, he was doing a man's day of work with the threshing crews, and he continued to work on the farm until he was twenty. He estimated he walked about ten thousand miles doing farm work.[18]

According to a E. C. Duffy, a member of the McCormick board of directors happened upon young Benjamin in the Iowa countryside as he was repairing a corn planter. After a few minutes of conversation, the executive was so impressed with the boy's abilities as a mechanic full of practical and common sense ideas on the subject of agricultural machinery that he hired him on the spot to come to Chicago and work for the company.[19]

The story is likely an embellishment at best, but it illustrates the lore that came to surround one of IH's most prolific and high-profile engineers.

Benjamin graduated from Iowa State College at Ames (now known as Iowa State University) in 1893 with a Bachelor of Science in mechanical engineering. While in college, he spent his summers teaching in country schools in Jasper County, Iowa.

However Benjamin was hired—and it's possible the roadside story is at least partially true—his first job was as a draftsman in the McCormick Harvesting Machine Company's experimental department. He advanced steadily throughout his tenure with that company and with IH after the merger, retiring in 1940 with more than 140 patents to his name. His patents include harvesting equipment as well as a wide variety of tractor advances ranging from shifting cultivator gangs and the drawbar to the seat spring and draft couplings. The power take-off developed at IH was his work, as was the improved McCormick twine knotter.

BERT R. BENJAMIN (1870–1969)

Credited as the father of the Farmall, Delbert "Bert" Rufus Benjamin came to McCormick in 1893 as a draftsman and ascended to become the superintendent of the experimental division at McCormick Works.

Bert Benjamin Archives

Benjamin was made superintendent of the McCormick Works experimental department in 1910, and in 1922 was named as assistant chief engineer in charge of Farmall development for IH.

Benjamin may have had a hand in the other farm tractor success story of first part of the century as well. He was sent to the Ford plant to cooperate on work on the Fordson in 1917. A war board decision intended to increase production put him there.[20]

Columnist Elmer Baker wrote about Benjamin in a 1959 issue of *Implement and Tractor*: "Then one man, not a committee, got an idea that a tractor that wouldn't displace the horses and mules on a farm was only half of the answer to the farm tractor. So he started to work on units that would

BENJAMIN FAMILY

Benjamin married his wife, Wilhelmina Louise Bergman Benjamin, on November 25, 1903. The Benjamin family came to America from England in the 1600s and were some of the first Europeans to settle in the United States. *Bert Benjamin Archives*

. . . cultivate. . . as well as doing the land locomotive work of other tractors.

"That man was Bert R. Benjamin, and he worked for the Harvester company experimental department. He may have had an office on the old seventh floor at 606 S. Michigan Ave., Chicago, but we always found him somewhere around that old shack close to McCormick Works next to the Illinois and Michigan Ship Canal."

In 1937, Benjamin left the corporate offices and was assigned to research work, which he continued until his retirement in 1939. Benjamin was all engineer, as he remained in hands-on positions to the end of his career.

The American Society of Agricultural Engineers honored him with the 1943 Cyrus Hall McCormick gold medal. The medal is awarded to an agricultural engineer by the McCormick family. Benjamin was only the twelfth person to receive this award. In a move of supreme irony, the fourth person to receive the medal was Theopolis Brown of John Deere. Much like Benjamin, Brown was a prolific engineer. Brown also filed a significant patent for a power lift similar to the one on the early Farmalls. IH was forced to pay Deere for Brown's power lift patent, and royalties were so significant that Brown kept detailed records of the earnings in his journal.

The expensive patent issue led the IH team to refer to Brown as "The Awfulest Brown." Giving him the Cyrus McCormick Medal shows that the family members awarding it had broad enough minds to look past the conflict and recognize Brown's accomplishments, and perhaps had a sense of humor.

E. A. Johnston was the recipient a few years later—and he certainly is one of the other stand-out engineers in early IH history.

In 1968, Iowa State University's College of Engineering awarded Benjamin the Professional Achievement Citation in Engineering.

Benjamin stayed spry and lively late in life. At ninety-three, he was still tackling new innovations, focusing his energy on researching how tire pressures related to improving automotive traction and safety, as well as regularly performing an exercise he had started when he was sixteen years old—whirling long-handled Native American clubs above his head.[21] Benjamin died in October of 1969 at age ninety-nine.

EXPERIMENTAL IMPLEMENT

While the Farmall was the product of many people, much of the Farmall development was done under Benjamin's watch, and he was an ardent proponent of the machine. *Bert Benjamin Archives*

IMPLEMENT DEVELOPMENT

Benjamin is best known for his work on the Farmall, but his more than one hundred patents include dozens of innovations for harvesting and tillage equipment. *Bert Benjamin Archives*

THE EARLIEST FARMALL DEVELOPMENT (1919–1920)

As chronicled in the beginning of this chapter, early development of the Farmall came in fits and starts, kicked off by Mr. Steward's acerbic observations at the large display of tractors at the 1910 Minnesota State Fair. Johnston's team were sketching heavily in 1912 and into 1914, and they even built an early prototype for a general-purpose tractor in 1915.

The line between the first Farmall and the last Motor Cultivator is a tad blurry. Whatever you call these early machines, starting in 1919 and 1920, the IH engineering team experimented with a reversible variation of a motor cultivator.

"There was talk about a new kind of tractor in the industry, but no one had such a machine or even much of an idea on how to start building one," Bert R. Benjamin recalled. "I knew we had to come up with an all-purpose tractor—one with rear wheels that could be adjusted to straddle two rows of crops and a narrow front with a wheel that would fit into one crop row."[22]

Benjamin said, "Working with [David] B. Baker, then chief engineer, and John Anthony, who took care of the drafting, we had an operating experimental model within three years."

So began the evolution of the Farmall. In 1920, the drive wheels were shifted from the narrow to the wide portion of the tractor. This move was probably prompted by Legge's insistence that the machine, "at least look like a tractor." This version also featured the automatic differential brake that was used on the Farmall. Note also that the "front" and "rear" of the tractor was a slippery concept; the 1920 version sported

1920 FARMALL "LIGHTWEIGHT" EXPERIMENTAL

Johnston noted this photo depicts the first experimental to use the final Farmall layout—large wheels in the back and the engine in-line. This model was reversible. Note the receptacle for a hand crank to start the tractor on the frame rail, just behind and below the radiator. Johnston stated he and Baker designed this in fall 1919, and the image was taken in November 1920. *Wisconsin Historical Society / 161139* three forward and three reverse gears and a reversible seat so that it could run either way.

This machine was known as the "cultivating tractor," and late 1920 photographs show versions of the machine bearing the name, "Farm-all." The tractor was reviewed in an IH memorandum as too cumbersome, too heavy, the steering action slow, and too difficult to get sufficient gang shift.[23]

The 1920 Farmall had some serious teething problems to solve, but Benjamin was still a firm believer in the basic design and concept. IH executives were divided on the topic, and the upcoming release of the all-new McCormick-Deerings did not help Benjamin's cause.

Benjamin provided evidence of the Farmall concept's viability by comparing the proposed Farmall 8-16 to the Moline Universal and the standard International 8-16. Each machine was evaluated on its ability to operate eleven commonly used pieces of machinery.

The Universal operated nine of the machines alone. The International 8-16 performed four operations without horses or additional persons. The Farmall could operate all eleven implements with only one operator. This was the true strength of the Farmall, and Benjamin had to play to it strongly to keep the project alive.

A film of a Farmall operating a ten-foot shocker and binder at Harvester Farm was shown to IH executives on December 13, 1920.[24] Immediately afterward, the group met to discuss the fate of the Farmall program for 1921. Most were unenthusiastic about the tractor. Despite its potential as an all-purpose machine, it was pointed out that the tractor and all the accompanying implements would be a sizable investment and that using the array of attachments would be complex for the average farmer. Also, the design was too heavy. Naturally, the failure of the Motor Cultivator was brought up as another negative aspect of the project.

Despite the fact that most IH executives had little faith in the project, it was determined that further exploration of Benjamin's concept for a general-purpose tractor was worthwhile. Five experimental models were authorized to be built in 1921, at a cost of over $150,000 because each example would be built by hand. Benjamin and a handful of engineers were reportedly the only people in the company with much enthusiasm left for the Farmall.

THE "LIGHTWEIGHT" FARMALL

Another view of the lighter Farmall shows the chain final drive, which is similar to the one used on the Tractor Cultivator of 1920. The chain drive was replaced with an enclosed gear drive some time in 1922. Also note the offset driver position—this idea would see production on the Model B. *Wisconsin Historical Society / 161140*

EXPERIMENTAL CORN CULTIVATOR 1920

The Tractor Cultivator was used to develop all kinds of interesting equipment, such as this No. 77A self-powered two-row corn picker. *Wisconsin Historical Society / 24697*

REVERSING DIRECTIONS TO FIGHT THE FORDSON (1921)

In January 1921, the authorization for five tractors was changed to two lightweight versions of the Farmall. The Farmall of 1920 weighed about 4,000 pounds, while the new prototypes were lighter.

At the same time, the cultivator was changed to mount to the front of the tractor. According to Arnold Johnson, an engineer working on Farmall attachments, the switch to a front-mounted cultivator went over well with his department and gained favor for the Farmall concept.

Of note is that the Farmall development was spread out over many departments. On July 21, 1921, the executive committee met and distributed the engineering work. The experimental department at McCormick Works was to build the Farmall attachments. The primary development of the Farmall Tractor was to take place at Tractor Works.[25]

Benjamin kept a steady stream of information flowing to Legge. A document was prepared comparing costs between a farm using eight horses, six horses and a Fordson, and a Farmall and two horses. The figures impressed Legge, and he passed the document around the company.

Benjamin then proposed a version of the Fordson tractor with the front wheels close together and the ground clearance raised in the rear. Such a machine could plow adequately and also cultivate. Benjamin felt sure it would reach the market, and that it was only a matter of whether IH or Ford built it first.

In the meantime, the company called a special conference to discuss what could be done about Ford. The Fordson was killing IH in sales, and solutions were hard to find. The Farmall program was discussed as a possible solution, but Alexander Legge sharply criticized the tractor's development. Johnston jumped to the Farmall's defense, and a chorus of lukewarm support followed. Some felt the Farmall could be developed into a better machine than the Fordson, and most thought there was promise in the idea. The result was that one hundred Farmalls were ordered to be built for 1922 (that would later be reduced to twenty). Also, a new line of implements was to be developed for the McCormick-Deering machines.

HEAVY FARMALL 1921

The reversible feature of the early Farmall prototypes was slick and allowed engineers to develop a wide array of front-mounted implements. *Wisconsin Historical Society / 24211*

The Farmall design continued to evolve. The gangs on the cultivator were designed so that they shifted when the front wheel turned, allowing the operator to easily follow the rows when cultivating.

Sometime in 1921, the chain drive was abandoned in favor of an enclosed gear drive similar to the type used on the McCormick-Deering tractors. By the close of 1921, demonstrations of prototype Farmalls had earned the tractor some grudging respect but not much interest from the sales department. Legge stated development was too slow.

Although Legge made scathing comments in a memo about the Farmall, he also made sure Benjamin received enough support. Times were tough for the general-purpose tractor program in 1921, and Legge knew that Benjamin was feeling down. Legge told a fellow employee, "Charlie, for heaven's sake go to Benjamin and see if you can give him some help and encouragement. He's got a real job on his hands and we want to see him succeed."

Legge believed in the concept—and in Benjamin.

Benjamin later recalled the time wryly. "They were ninety-nine to one against building this tractor," he said. "But I kept on with it—even when I didn't have the authority to do so."[26]

HEAVY FARMALL WITH CULTIVATOR

The experimental Farmall most favored in 1921 was heavier than the 1920 edition and had a longer frame.

Wisconsin Historical Society / 22765

HEAVY FARMALL

These prototypes saw the chain final drive replaced with an enclosed drop axle that was powered by a massive transmission and a final drive that housed the power lift, power take-off, and belt pulley.

Wisconsin Historical Society / 161234

THE BATTLE TO FORM THE FARMALL (1922)

In January 1922, the company authorized twenty Farmalls to be built. Actual production was seventeen machines, which were heavy, reversible examples.

Legge remained skeptical of the Farmall. In a memo to a fellow executive, Legge wrote, "About the only thing we have settled so far is that we have done a very poor job of putting [Farmalls] together, which suggests that you should try to strengthen your engineering staff."

In June, Legge said that the process of having IH representatives closely guide farmers who were testing experimentals skewed the feedback. His concern was if the machines were so complex that the only way to operate them was with expert guidance, they were a long way from ready for production.

In July, Benjamin demonstrated a lighter version of the Farmall at the Hinsdale Farm. The machine showed enough improvement from previous versions to keep the project alive, and "Mr. Legge said that after seeing this tractor at work at Hinsdale he was at least satisfied that some progress was

HEAVY FARMALL 1922

Development continued on this prototype into 1922. Note that the fuel tank moved on various models—some had a tank over the engine, others, as shown here, had a round tank mounted above the steering gear. *Wisconsin Historical Society / 161179*

IMPROVED LIGHTWEIGHT FARMALL 1922

This radical redesign of the Farmall was sketched by Bert Benjamin in mid-1921, blending features of the early experimentals—particularly the ability to cultivate—in a clean, sleek design. *Wisconsin Historical Society / 161235*

IMPROVED LIGHTWEIGHT FARMALL 1922

This experimental—also called the high frame or high side-rail model—was demonstrated at Hinsdale in July 1922 to Alexander Legge and others. The machine performed well and convinced IH management that the Farmall had a future. *Wisconsin Historical Society / 161231*

being made." Before Legge left Hinsdale, the original order for twenty Farmalls was canceled, and the lighter version Benjamin demonstrated was to be developed further with two units built.

Benjamin filed a memo in February 1922 that compared costs to farm 120 acres of land that had 40 acres of corn, 20 acres of oats, 30 acres of hay, and 30 acres of pasture. This was the size of 60–80 percent of American farms. He showed the hours required to work that land for five horses was 122 days of labor, but with a Fordson and two horses it required 108 days of labor. The Farmall 6-12 could do the work in just eighty-eight days.[27]

Also in 1922, Benjamin wrote a prescient memo predicting that the Farmall would replace the horse. "I predicted in 1922 that two million tractors would take the place of 20 million horses and mules," Benjamin said in 1963. "I said that this would save the farmer one billion dollars' worth of fuel, labor, and feed. Maybe history has proved me right."[28]

In November 1922, Benjamin was transferred to the "city office" and assigned to coordinate all the work in connection with the Farmall. Prior to this, the coordination had been done by McCormick Works people—David B. Baker, Philo H. Danly, and John Anthony.[29]

LIMITED RELEASE (1923)

The year began well for the Farmall—with a caveat. Legge kicked it off in February 1923 when he presented his thoughts on the Farmall to the executive committee. He traced the evolution of the machine, and shared with his fellow IH leaders that he believed the team was close to coming to an agreement on the form of the machine, and that what they had would meet the objectives of the wide variety of stakeholders in the project.

Some members of the leadership team felt two tractors were needed to perform farm tasks—a cultivating/harvesting machine and a plowing machine. Others had concerns the Farmall release would harm sales of the McCormick-Deering 10-20.

Meanwhile, the engineering team continued to debate the details. While the layout had been established in late 1922, the 1923 Farmall discussions centered on power, weight, and cost. The team felt that the tractor needed to be capable of pulling two 12-inch plow bottoms and needed to weigh about 3,000 pounds. Cutting back features and quality to reduce cost was seen as a mistake. Note that the weight of the Farmall would creep to just over 4,000 pounds by 1925.

The manufacturing arm of IH was occupied with building McCormick-Deering tractors, so the engineering department would have to construct the 1923 machines. It was decided that the tractors needed to be capable of pulling existing implements as well as custom-made units. A maximum of twenty-five of

1923 FARMALL EXPERIMENTAL

This photo dated April 1923 is of an experimental Farmall that is very similar to early production models. The shifting cultivator gangs shown on this example were a key feature to the tractor's eventual success. *Wisconsin Historical Society / 161132* the "lighter-weight" Farmalls were authorized to be built in 1923, along with twenty-five cultivating attachments. The specifications for the machines were essentially the same as those of the 1922 machine, with the exception of trimming about 300 pounds of weight. The weight was approximately 3,229 pounds.

The first of these tractors were sent to Insull's farm in Libertyville, Illinois, and the machines were observed continuously by Benjamin and John Anthony. Two of these machines were used there for several years, logging a total of fifteen thousand hours.[30]

In April, Legge wrote Johnston, stating his concerns that the cost of the McCormick-Deering and the Farmall were too close. This was an ongoing debate that centered around the question of whether farmers would need two tractors—one to plow, and one to cultivate—or one that could do both. While we now know that a Farmall could ably perform both tasks, that was not widely understood even as late as 1923.

Reading between the lines, Legge was reflecting the executive committee's position that the Farmall should be a secondary purchase, and therefore significantly less expensive than the McCormick-Deering. At the same time, Legge was discouraged to discover that the engineering department was developing several types of rear gear drive. He felt the tractor had been developed enough, and that there was not time nor resources for drastic engineering changes.

Legge wrote, "I think perhaps it is true that in the past we have brought out some of our designs a little too hastily; however, this is not true of the Farmall, as you have been building it up and down for five or six years and by this time ought to have a pretty fair judgment as to what is the best and cheapest plan to follow."[31]

The new development was a revised housing for the rear end, and Legge's letter put an end to that.

The Farmall name was registered July 17, 1923, and authorized by the naming committee February 5, 1924.

According to Farmall Tractor Works booklet, twenty-two Farmalls were built in 1923.[32] Page 102 states twenty-six were built in 1923—and thirteen were experimental models. Fascinatingly, several of those 1923 Farmall experimental models went to the farms of IH executives. One went to Harold Utley's farm, and another to the Cyrus McCormick Jr. farm.

FARMALL EXPERIMENTAL JUNE 18, 1923

IH manufacturing was at capacity building McCormick-Deering tractors, so the engineering team hand-built their prototypes for 1923. Fewer than two dozen of these were built and tested. *Wisconsin Historical Society / 161181*

RAVE REVIEWS

A September 29, 1923, report in *History of a Tractor* (Seyfarth, 103) on experimental Farmalls placed on nine farms in Illinois and Wisconsin was "universally favorable . . . particularly the ease with which the tractor was handled." *Wisconsin Historical Society / 161384*

THE FIRST TWO HUNDRED FARMALLS (1924)

The tractor was strengthened in an assortment of key areas for 1924. The bolster became a single casting, the bull gear housing was strengthened, and the rear axle was increased in diameter. The drawbar fastening was enlarged, the rear axle roller bearings were equipped with an oiler, and fuel capacity was increased to thirteen gallons.

The production run for 1924 was limited to two hundred units, with the price below cost at $825 (they would later adjust the price to $950). The numbers were kept small to avoid hurting sales of the similarly-priced McCormick-Deering 10-20. The tractors were sold through dealerships, and each model was closely monitored to determine performance and reliability.

The first tractor out the door was QC501, which went to Taft, Texas, along with the first skeleton wheels. Benjamin visited this machine for several weeks in 1924 and for several years after that. The machine performed well, harvesting cotton with only minor maintenance required.

One of the most significant Farmall advances appears to originate with QC501. When Benjamin visited that machine, he became convinced that IH should build a four-row cultivator. He stated this in a June 22, 1924, letter to the IH patent department.

The four-row cultivator was part of the magic of the Farmall, as it made the farmer more productive. The major deficiency of John Deere's ill-fated Model GP, which was developed as a response to the Farmall, was to use only three rows—one of the few misses in the illustrious career of Benjamin's green machine counterpart, the brilliant Theo Brown.

1924 FARMALL

Serial numbers for the 1924 Farmalls started with QC501, and just over two hundred were built, with a little over half sold and the rest used as test and demonstration units. This image is dated March 5, 1924. *Wisconsin Historical Society / 161131*

1924 FARMALL

The distinctive visual features of a 1924 Farmall are the tall intake and exhaust pipes, with the cotton air cleaner on top of the intake. The 1924 models were built on sawhorses in a small corner of Tractor Works. This image is dated March 5, 1924. *Wisconsin Historical Society / 161129*

1924 FARMALL AT WORK

As 1924 closed, the reviews of the Farmall had the IH management team conservatively increasing build numbers and projections. Edward A. Johnston believed the only way to compete with the Fordson—which held more than half the tractor market in 1924—was to offer a machine with greater utility to justify a higher price. "In his judgment, the Farmall was the answer." (Seyfarth, 105). This early production model is shown in an image dated June 10, 1924. *Wisconsin Historical Society / 22777*

FARMALL QC503

The tractor in this image is most likely the first Farmall sold, QC503, which went to a farm managed by Roy Murphy and was used there for nine years. This photo dated August 21, 1924, is captioned "J. F. Deems Farm, Roy Murphy Manager, Burlington, Iowa." IH later purchased the tractor, and QC503 remains in the company collection. *Wisconsin Historical Society / 120981*

Benjamin felt the four-row cultivator would save ten cents per acre. He would write several letters to this effect, along with much more technical detail about how the cultivators would need to function. This innovation would greatly improve the Farmall's utility.

The first Farmall sold—according to company statements—was serial number QC503, and it went to Roy Murphy of West Burlington, Iowa. The tractor was delivered on March 1, 1924.[33]

In the showroom, the tractor had some other hurdles to overcome. The tricycle design was unusual at the time, and the Farmall was rather gangly and unfinished when compared to the compact, more graceful lines of the popular standard tractors like the McCormick-Deerings.

In the field, however, the testimonials were consistently positive. Legge kept a file of these showing "the Farmalls put out in 1924 were giving a good account of themselves." One Texas cotton grower said, "It's homely as the Devil, but if you don't want to buy one you'd better stay off the seat."

The tractor sold for $950 in 1924; 205 were built and sold, with nine of those delivered to Texas. Throughout 1924, the tractor's performance was lauded. Company executives believed the tractor would probably kill the McCormick-Deering 10-20, but it also felt that it would be better for an IH product (rather than a competitor's) to knock off the 10-20.

In September 1924, Edward A. Johnston declared that there is "a constant cry to for a tractor to meet the Fordson." In his opinion, the Fordson price was so low that the only way to beat it was with a machine that offered better utility. He believed that the Farmall in 1924 was just that machine.[34]

"THE FIRST FARMALL"
This display was likely at the 100 Years of Chicago celebration held in 1947. IH had several early Farmalls—this is almost certainly QC503. After it was sold back to IH, it was fitted with odd decals (as shown). Case IH had it properly restored much later. *Wisconsin Historical Society / 59370*

FARMALL WITH TWO CORN PLANTERS
This 1924 model was photographed on the University of Illinois farm. Caption notes that the Farmall is capable of working with two corn planters. The Farmall's horsepower ratings were deliberately not listed or advertised—company literature offered that the tractor could pull two 14-inch bottoms in average condition. This photo is dated May 1925. *Wisconsin Historical Society / 161233*

THE TIDES TURN (1925)

As the Farmalls flowed into the world, the feedback coming back to IH was starting to overwhelm corporate resistance to the concept. The machine not only functioned fantastically, but it was also a well-built tractor. Experimental machine QC294 ran the equivalent of twelve thousand miles over four hundred days in Corpus Christi, Texas, for farmer C. C. Bolte with no visible wear on the main gears and bearings.

In the region, the Farmall was making good impressions. Corpus Christi-based engineering department leader J. J. Hipple wrote, "The Farmall is becoming very popular, and they all want to use them for planting as well as preparing the land." He added that farmers in the region were able to hook two two-row planters to the Farmall with good results. One farmer told him he planned to replace his mules with "at least" ten Farmall tractors.

Ever the mathematician, Benjamin used March 1925 findings at the J. F. Jones cotton farm to calculate that three hundred thousand Farmalls could replace the 2.5 million mules used for cotton farming worldwide.

For 1925, the Farmall was again improved. The changes were fairly slight but covered most of the tractor. A host of systems were simplified, including the gear shift, starting crank, steering gear, front bolster assembly, differential assembly, transmission assembly, and even the toolbox. In addition, the wheel hubs and transmission case were strengthened, and a muffler and air cleaner were added. The tractor weighed just over 4,000 pounds. International management's reluctance to put the Farmall into production resulted in a better machine. With each delay, the tractor was refined and improved.

Although only 250 were originally authorized to be built in 1925, the demand prompted IH to build 838 Farmalls that year. Many were sent to Texas and sold for $100 more than the 10-20. It was felt that the tractors would sell to cotton growers, opening the market to IH without hurting 10-20 sales. Interestingly, IH did not devote an advertising budget to the Farmall. This move reflected the company's desire to avoid interfering with McCormick-Deering sales. The publications of the time reflect the decision, as Farmall ads are scarce and IH's magazine hardly mentioned the Farmall until 1926 (in fact, IH did not dedicate any national advertising to the Farmall in 1925).[35]

Part of the IH strategy in 1925 was to release the machines into territories that had very little IH tractor business. Nebraska, in particular, was selected as one of those places, and several demonstrations were held in 1925 to help move the machines. The south and southwest were also targeted. In the central Midwest, Farmall sales were curtailed, with specific instructions from top brass that under no circumstances were Farmall sales to interfere with sales of the McCormick-Deerings.[36]

The reports were again favorable on the 1925 machines. In a company memo, G. A. Newgent responded to queries about how the Farmall was performing in Texas. Rather than reply in his own words, he gathered a few statements from farmers who were using the new machine. They included the following:

"Well, I am buying my second Farmall today. When I bought the first last spring I was afraid it would give me considerable trouble as it looked spiderly, but there has not been a cent spent on it and it pulls as well today as it did the day it was delivered to me."

As well as this gem:

"My wife uses a nickel's worth of ice per day and the drippings from the ice box are more than ample to water my Farmall."

FARMALL EARLY PRODUCTION

This image is dated November 28, 1924, meaning this is a 1925 model (IH started its production of the following year's machines on November 1). In 1925, IH had to settle a patent dispute on the Farmall cultivator design with Adolph Ronning. Rather than risk a lawsuit, IH paid Ronning a $25,000 advance and $1 per Farmall sold after May 1, 1926. *Wisconsin Historical Society / 161133*

Newgent went on to write that the 250 Farmall owners were, without exception, extremely pleased with the Farmalls. He closed with this: “When a farmer buys a Farmall he joins the sales force of the International Harvester Company.”

That was strong testimony, but another Texas sales rep had stronger testimony to offer. Several general office men—including representatives from sales, manufacturing, and engineer Bert Benjamin—met with a group of International’s Texas salesmen. The Texas representatives wanted the company to build the Farmall for regular production. P. Y. Timmons, the general office manager of tractor sales, said that it might affect sales of the McCormick-Deering 10-20. The response was that the Texas branches didn’t sell many 10-20s. J. F. Jones, general office sales manager, commented that if anyone was going to build a tractor that would displace the 10-20, it might as well be Harvester. Lastly, Jim Ryan, the manager at Houston, said, “If you don’t adopt it for production, we will organize a company in Houston and build it down here.”

This kind of statement was hard to ignore. Even so, the concern over producing a tractor that competed with the McCormick-Deering persisted. In December 1925, Legge wrote that, despite concerns, the Farmall production for 1926 would be increased.[37]

1925 FARMALL ADVERTISEMENT

In 1925, Farmall advertising was limited to regional (not national) coverage, as IH management was concerned about slowing sales of the McCormick-Deering 10-20. The early Farmall paint scheme featured red wheels and a gray body, with other colors available only by special order. *Wisconsin Historical Society / 4341*

BUILDING FARMALLS

For 1925, 870 Farmalls were built at Tractor Works in Chicago. IH purchased the Moline Plow Company factory in Rock Island, Illinois, in 1924, renamed it Farmall Works, and transferred Farmall production there in October 1926. This photo was taken inside Farmall Works in April 1928. The Farmall Regular, F-20, and F-30 were built at Farmall Works in Rock Island. The F-12 was built at Tractor Works in Chicago. *Wisconsin Historical Society / 47007*

ON THE BANDWAGON (1926–1932)

In 1926, the Farmall's sales began to take off, and IH was finally backing the Farmall concept. A new plant was opened in Rock Island, Illinois, specifically to build Farmall tractors. The plant was known as Farmall Works and, once it got started, tractors rolled out the doors at an incredible rate.

The machine received a generous splash of attention in *Tractor Farming*, IH's promotional magazine. The Farmall was on the cover of the March-April 1926 issue, and the magazine lauded the merits of the company's "new" machine.

"For the last three years, this new tractor has been proving its mettle and merit in considerable numbers in a few limited sections, primarily the Southwest cotton territory.

"Quantity production is now getting under way, though it will be some time even yet before production can overtake the demand, especially now that the Farmall will be supplied, on order, to any section of the country."

The 1926 machine was improved again, and it certainly benefited from the long development phase. Each year, a few new problem areas were identified and strengthened. The result was that, by 1926, the Farmall was a highly refined machine.

IH also marketed the Farmalls in the earliest days very carefully, pushing the machines hardest in regions of the country where IH products—particularly smaller tractors—did not sell well.

Sales exploded to nearly twenty-five thousand in 1928, and the Farmall was off to the races. It did just as well as IH management had hoped. Also, the McCormick-Deering 10-20 sales proved much more resilient than feared. Perhaps due to IH's selective market choice, perhaps because plenty of farmers preferred the traditional looks, the McCormick-Deering models sold very well.

TRAINLOADS

By 1926, Farmall demand accelerated and 4,400 were produced. Production at the new Farmall Works increased radically, and, by the time this image was taken in July 1928, Farmall Works was building 125 tractors per day. *Wisconsin Historical Society / 133283*

1927 FARMALL

The decision to replace the stenciled Farmall logo on early Farmalls was made on March 8, 1926. The Farmall would evolve during production, with small changes throughout the years. This image is dated December 22, 1927. *Wisconsin Historical Society / 24700*

1929 STATE FAIR

The Farmall was nimble and easy to drive compared to the traditional standard tread tractors of the day. It turned on a dime, partially thanks to a cable system that activated the inside rear brake when turning sharply. This would spin the tractor around on the rear wheel. This image shows a Regular on display at the 1929 Indiana State Fair. *Wisconsin Historical Society / 52212*

1927 FARMALL REGULAR

This Regular spent seventy-three years on a ranch near Boerne, Texas, before being purchased by Harold Sohner in 2000. This 1927 production year has special significance, as Ezra Sohner, Harold's uncle, took the role of general superintendent of Farmall Works in 1925. In addition, Harold's father, Harold K. Sohner, began work at the same place in 1927, meaning two Sohner family members were at Farmall Works when the tractor above was built. *Lee Klancher*

FAIR SHOWCASE

By 1929, the Farmall was headline news for IH. The Fordson sales had died, and in fact stopped production completely, at least in part due to the popularity of the Farmall. This Regular with a dubiously safe saw attachment is on display at the Mid-South Fair held in Memphis, Tennessee, in 1929. *Wisconsin Historical Society / 52209*

FEDERAL FARM BOARD 1929

IH lost a key executive when Alexander Legge (standing) was called to serve as chairman of the Federal Farm Board in 1929. He did so with honor, and again showed the world what an uneducated kid from Nebraska with a good head on his shoulders could accomplish. *Library of Congress LC-H2- B-3591*

EXPERIMENTAL COTTON PICKER

One of the high-priority developments of the 1920s was the cotton picker. IH was one of the pioneers, and testing was done on Farmalls. This photograph is dated March 16, 1927. *Wisconsin Historical Society / 24586*

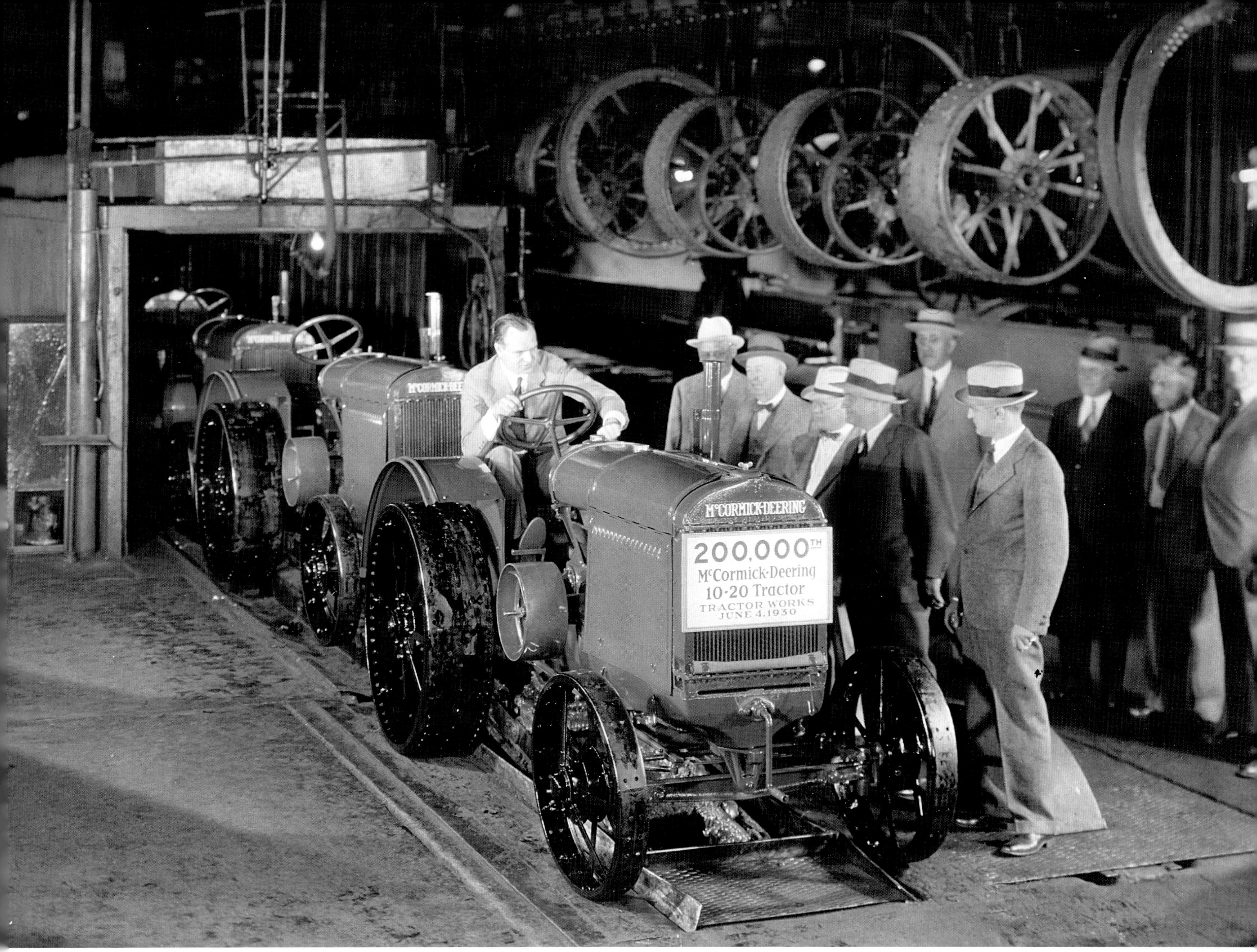

200,000TH MCCORMICK-DEERING 10-20

Fears that the Farmall would interfere with McCormick-Deering sales proved unfounded. The historic 200,000th McCormick-Deering 10-20 was driven off the line at Tractor Works on June 4, 1930, by Cyrus McCormick Jr. *Wisconsin Historical Society / 7241*

FARMALL FAIRWAY

In February 1929, IH management decided to offer a version of the Regular designed for the golf course. The earliest examples had "Fairway" stenciled on the fuel tank in place of the "Farmall" decal. Later models had a new decal with the word "Fairway," as shown on this image dated 1930 and taken at the Chariton Golf Club near Chariton, Iowa. *Wisconsin Historical Society / 6973*

NARROW-TREAD FARMALL

A narrow-tread version of the Farmall was built to accommodate farmers who required a narrower wheel set. According to IH documentation, the need for such a tractor was especially great in Argentina, so these tractors are probably quite rare in America. The tractor was equipped with an offset rear hub that narrowed the rear tread width from 80 inches to just over 63 inches. The narrow tread was authorized in the spring of 1927. A variety of wheels with different offsets were offered to allow for different tread widths.

Another special edition of the Farmall Narrow-Tread was the Farmall Fairway, which was designed for use on golf courses. Smooth wheels were used, as well as offset wheels that narrowed the tread width to 57 inches. In 1929, a set of lugs was authorized for the golf course machines, probably due to wheel-spin on wet grass or hillsides. The lugs were simply small spikes.

Perhaps the rarest of the Farmall attachment packages was the orchard fenders. These were available from 1931 to 1934, and then were dropped due to lack of demand. This is easy to believe, as the ideal orchard tractor is low-slung, which the Farmall is not.

NARROW-TREAD FARMALL F-30

IH developed narrow-tread versions of the Regular as well as later models like this F-30. While the standard Farmall wheel width could be altered with different wheel offsets, a narrower version was developed and offered after May 5, 1927. The standard tread width was 74 inches, and the narrow-tread model was 57 inches wide. This image is dated January 28, 1932. *Wisconsin Historical Society / 46665*

In 1930, a field change was supplied for early 1925 model Farmalls (serial numbers QC700 to QC8821). The change package improved the tractors' durability, especially in dusty conditions. Most of the changes centered around better sealing against dirt. It was authorized in February 1930.

THE 100,000TH FARMALL

When IH produced its 100,000th Farmall, it painted that machine bright silver and celebrated it rolling off the production line at Farmall Works on April 12, 1930. The milestone was all the more remarkable considering the fact that only a small portion were built from 1922 to 1927. The bulk of the machines were built and sold from 1928 to 1930.

The event was attended by much of IH top management, including president Harold McCormick and Cyrus Jr., who drove the machine off the line and gave the celebratory address.

They also brought in the company's "first Farmall," serial number QC503, to show off the progress from the first example to the shiny latest entry.

The machine lives on today due to the efforts of Ryan Peters, a collector from central Iowa. He purchased a number of old machines from a friend of his and, months later, was examining the rusty treasures when he found a Farmall Regular with "100501" stamped on the frame rails and the engine of one of his old machines. Peters did his research and found that the machine was IH's 100,000th Farmall tractor. He found remnants of silver paint under the hood, on the gas tank, and on the hand crank. He decided to restore the rare tractor to its original glory. The mechanicals were in pretty good shape, with low wear on the steering gears and drawbar, indicating the tractor hadn't worked much.

In the original article about the 100,000th Farmall in *Harvester World*, there is brief mention of the tractor's destination: "the 100,000th Farmall went off the line and into the world of work and a useful life, or, as was later suggested, to be permanently displayed as a subject of historical interest."

1930 100,000TH FARMALL
This rare machine was found in a pasture and restored by Ryan Peters. The hand crank remains unpainted, with the old flecks of silver from the historic machine intact. *Lee Klancher*

OFF THE LINE
Cyrus McCormick Jr. is shown here riding the 100,000th Farmall tractor off the Farmall Works assembly line on April 12, 1930. The tractor was painted silver to mark the occasion. At the front of the line is Superintendent E. H. Sohner. This same tractor was found and restored by Ryan Peters. *Wisconsin Historical Society / 12105*

100K CELEBRATION
Several executives spoke at the celebration of the 100,000th Farmall production, including Cyrus McCormick Jr., shown here on stage. IH also had the 99,999th tractor on display. *Wisconsin Historical Society / 25912*

THE FINAL DAYS FOR THE REGULAR

According to most documentation, production of the original Farmall ended in 1932. But an assortment of company documents indicate that a limited number of Farmalls were sold in 1933 and 1934. The most convincing of these is the record of payments to the Ronning family. The Ronnings owned a patent to several features similar to those of the Farmall, and IH had to pay the Ronning family one dollar for every Farmall Regular, F-30, F-20, F-12, and F-14 sold.

Records of the payments for 1933 and 1934 show several hundred dollars paid for Farmall Regulars produced and sold. More likely than not, a few tractors from the 1932 production run were left over and sold in 1933 and 1934.

Near the end of the Farmall's production run, IH authorized optional pneumatic tires for its tractor lineup. Although the decision wasn't signed until 1933, after the Regular was out of production, it listed 9x36 rear and 6x16 front pneumatic tires as additions for the Farmall. It is doubtful that many Farmall Regulars were equipped with these tires from the factory, but dealers and owners almost certainly retrofitted pneumatic tires. The attachment number for the rear tires for the Regular was 25711-D; the number for the front tires was 25494-D.

A later retrofit was a hydraulic lift, which was authorized in 1945. The lift was a version of the new Lift-All adapted to Farmall Regular, F-20, and F-30 tractors. The part numbers were 350420R91 for the Regular, 350421R91 for the F-20, and 350422R91 for the F-30.

The Farmall design proved to have lasting merit, and the Farmall formed the basis of the IH lineup for the next several decades. The concept came to life through a number of avenues. Inspiration came from the observations of the cranky old-timer, John F. Steward, of the Hackney tractor in 1910. Then Johnston and his engineering team spent several years drawing and building the new model.

The idea truly blossomed under the tutelage of Benjamin, who received a mix of encouragement, guidance, and funding from the Nebraska cowpoke-turned-collector-turned-leader-of-IH, Alexander Legge.

There are others who contributed, but Benjamin was the mind who guided and drove the Farmall design to succeed and become the first all-purpose tractor. The road to this success was filled with potholes and roadblocks.

C. W. Gray claims the resistance to the Farmall was as much political as it was practical. Benjamin was often the only proponent for the Farmall, and Gray states that Benjamin threatened to leave unless the project received the funding and support he felt it deserved.

The giant merger that formed International Harvester is many different things to many different people. While it is not the cruelest or most punitive of the era's mergers, the fact that so much of the industry went under one roof made business harder for others. Did the consumer—the farmer—suffer at the hands of IH? It appears not, but the new trust's labor practices were problematic, and some of its business practices questionable.

What cannot be questioned is that one of the great innovations of the twentieth century emerged from this company, and those who created it came from several different arms of the many-headed beast that was IH.

1930 FARMALL REGULAR

This original condition Regular is part of the Jon Kinzenbaw collection. *Lee Klancher*

FARMALL LEGACY

The Farmall Regular proved to be one of the most important tractors of the twentieth century, and the easy-to-use little tractors would prove useful on farms for decades. This one is shown working the fields on a sawmill run by a five-family co-op in Gem County, Idaho. *Dorothea Lange, Library of Congress LC-DIG-FSA-8b35410*

History owes a tip of the hat to Benjamin and the Farmall, as both ushered in a new era of farm power. The horse, once vital to farming, was gradually replaced with the tractor, and the Farmall was the first tractor able to fill the horse's shoes. Some debated the merits of replacing flesh with metal, a companion animal with a cold steel beast, but it was impossible to debate the merits of increased productivity and decreased manual labor. The general-purpose tractor had arrived.

PLOWING AHEAD

Fowler, Indiana. in April 1937. *Russell Lee, Library of Congress LC-DIG-fsa-8b30310*

CHAPTER FOUR

POWERING THROUGH HARD TIMES

1931–1938

"[The] tractor does two things—it turns the land and turns us off the land. There is little difference between this tractor and a tank."
—John Steinbeck, *The Grapes of Wrath*

During harvest season 1929, Lawrence Svobida was up long before light to grease and repair his tractor and combine. He had hired help for the season, but their time was dedicated to bringing in crops from his Kansas wheat fields. Mechanically inclined and hardworking, Svobida got up well before the sun to personally attend to his machines. The gritty mechanical work often took two or three hours, and he would frequently work into the wee hours in his struggle to make the farm a viable business.[1]

Svobida came to Meade County, Kansas at age twenty-one, a strapping young man who had already proven his mettle as a champion corn shucker capable of husking as much corn as any two men in the county. He took over his father's land, confident that his work ethic could turn the fertile ground into a successful venture.

The times were dark; America was plunged into the Great Depression when banks failed, and the worldwide economy took the deepest dive in history. Svobida was young, strong, and confident—a shining example of the optimism and grit that characterized early twentieth century America. His outlook was brightened by the rise of power farming—the new machines offered the ability to plant and harvest at a scale his father could have only dreamt about.

Svobida's generation of farmers saw that the advent of new technology—specifically tractors—would improve productivity on the farm, which in turn would allow a relatively smaller population of farmers to produce enough crops to meet the demands of the world's rapidly expanding population. The first World War drove the demand for crops worldwide, and the depletion of the workforce from men going to war encouraged the use of these new technologies. When the war ended and the men returned, they employed tractors and improved implements to open up ground at a rate never before experienced.

But the power to transform native grasses into productive pasture had a destructive side as well. Like the atomic bomb, the Internet, and social media, power farming offered new capabilities with unforeseen dark consequences.

As farmers like Svobida flocked to create new farms from fertile ground, rapidly changing worldwide demand

DUST BOWL

A farmer raising his fence out of the dust in Cimarron County, Oklahoma, in 1936. *Arthur Rothstein, Library of Congress LC-USF34-004051-E* for agricultural products and decreasing prices drove them to produce more and more. With the new tractors at hand, hardworking farmers could open up thousands of acres. The new ground required no fertilizer and produced tremendous crops, at least for a short time.

After a few years—and under the pressure of drought and high winds—the rich and newly overturned dark soil turned to dust and literally blew away. Svobida's region held, at least for a time. As soil blew away in the Dakotas in 1929, Kansas farmers largely had respectable crop yields from 1928 to 1931.

This was the environment in which IH was able to sell its new Farmall and McCormick-Deering tractors, even as the economy nosedived in 1929. Given the ability to buy on credit and the need to produce, farmers could take a gamble and add the machines (and debt) to their operations while trying to produce enough crops to stay afloat.

The combination of new technology, falling prices, rising demand, and a global economic crisis would prove to be tragic for the farmer, creating the most difficult decade in history for farming and the country.

The tumultuous economy did not slow the need for bigger and better machinery, nor did it stop farm tractor makers from continuing to innovate and market. The Farmall had been out since 1924, and the rest of the tractor industry had taken notice and introduced general-purpose machines of their own.

International responded in the early 1930s with a formula that it would repeat for a century: the way to sell new machines is with more horsepower. The company introduced a new, high-horsepower version of the Farmall, the F-30, and an increased-power version of the McCormick-Deering, the W-30. They then added power and features to the Regular to create the F-20.

But a good harvest alone does not guarantee a farmer a decent yield. The plummeting prices of early 1931 required many farmers, including Svobida, to open more land in search of profit. He worked around the clock in his quest to earn a living on his farm.

His hired man ran the tractor throughout the day, and he would take over and personally drive the dangerous night hours. "Often I could hardly keep my eyes open," he wrote. "Only an instinctive and vicelike grip on the steering wheel kept me from falling and being ground to pieces by the plow . . . many a man has met his death this way, but I was lucky."

Farmers across the nation faced similar challenges, and machinery sales would plummet in the early 1930s. IH did not turn a profit in 1931 and chalked up significant losses in 1932 and 1933.[2]

By that time, the government was trying to help mitigate the disastrous economy and the farmers devastated by the situation. Federal loans offered Svobida a small solace, but the fact remained he had only successfully harvested one crop of the six he planted in his four years on his father's property.

International looked to provide machines that suited the times, and the little F-12 of 1932 did that well, offering the struggling farmers of the day a lower-cost entry into the high-tech world of power farming. International bookended that with its new, higher-horsepower W-40. The diesel version and large crawlers emerged in the mid-1930s as the farm economy

TRACTOR DRIVER
This former tenant farmer drove a tractor on his farm in Bell County, Texas. In June 1938, his pay was one dollar per day (twenty dollars in 2023 value). *Dorothea Lange, Library of Congress LC-DIG-fsa-8b32103*

experienced a mild rebound that put IH's books into the black as it recorded mild profits.

By 1935, the black blizzards of dust storms in the Great Plains turned day into night, highways into deserts, and infected healthy people with "dust pneumonia," which left them coughing up large chunks of dirt and often turned into tuberculosis. The dust storms often disabled an engine's electrical system, and the right mix of wind direction could fill an engine with dust and wedge the pistons in place. Tractors would become buried in the silt; wheels spun deep into the loose soil, and long poles had to be used to extract the machine.

COUNTY AGENT
Some Government programs during the Depression sent people to help solve technology issues on the farm. This county agent is helping a farmer adjust his plow in May 1940 in Jasper County, Iowa. *John Vachon, Library of Congress LC-USF34-060887-D*

With visibility at zero during these storms, crashes on highways were common. And if the vehicle occupants ran outside in terror—which happened too often—the dust storm suffocated them. The clouds would drift all the way to the Eastern seaboard that year, with some traveling as far as Washington, DC, where a variety of assistance programs were debated, drafted, and, at times, rescinded, with most providing some relief and none of them effective enough to turn the tides.

In 1937, Svobida reported a new threat—the stripped bare ground combined with hot sun and a stiff breeze would superheat the air. The super-heated winds burned his wheat crop that year "like a blast from a huge, red-hot furnace."[3]

Svobida continued to fight, bending his back during long days and using his innate mechanical ability and creative farm methods to produce enough to stay alive and assist his neighbors, particularly with repairing their aging and often poorly-maintained machinery.

"The owners of these neglected machines are doing the best that they can in their circumstances of financial stringency. They have no money and they can borrow none, so they make what repairs they can and hope against hope that their engines will hang together until a season's work is finished."[4]

Despite the grim conditions on most farms, IH sold enough machinery late in the 1930s to turn solid profits. They reduced employee wages by 27 percent from 1936 to 1937, a cost-cutting measure that would not help its contentious labor relations.[5]

In 1937, Svobida's equipment was aging as well, if at least slower than his neighbors' equipment. He rebuilt engines in winter blizzards, and sweated his clothes into pieces on his combine harvesting. With sleepless nights and working round the clock seven days a week, he was able to harvest six thousand bushels of grain from roughly 545 acres. He sold four thousand bushels for $1.10 each and held on to two thousand, hoping for higher prices. The harvest would keep him just solvent enough to stay on the farm.

He would not find better returns in 1938 and would give up in 1939. "With my financial resources at last exhausted and my health seriously, if not permanently impaired, I am at last ready to admit defeat and leave the Dust Bowl forever. With youth and ambition ground into the very dust itself, I can only drift with the tide."[6]

While Svobida and millions of others struggled to make their farms work through the end of the 1930s, IH had placed the sure bet that the farmers that remained after the struggles would need improved machinery. The company invested heavily in research and development and would close out the decade with the most salable machines built in recorded history.

FSA TRACTOR

This tractor was purchased with assistance from the Farm Security Administration and was photographed in Box Elder County, Utah, in August 1940. *Russell Lee, Library of Congress LC-USF34-037426-D*

ARID LANDS

A dust bowl farmer and his son working the dry ground near Cland, New Mexico, in June 1938. *Dorothea Lange, Library of Congress LC-DIG-fsa-8b32410*

FARMALL F-30

While the Fordson posed a major threat to IH, other manufacturers also provided challenges. IH enjoyed a short-lived moment of exclusivity with the general-purpose Farmall, but that was not likely to last. The tractor's genius was its simple tricycle configuration with large rear wheels. This was easy enough to duplicate, and the competition quickly followed suit.

The John Deere general-purpose tractor first appeared as the Model C in 1927 (and became the GP in 1928 and beyond)—timing that certainly could have spelled trouble for IH, given the fact it didn't gear up high-powered promotions of its new Farmall until roughly that same year.

Fortunately for IH, Deere had uncharacteristically rushed the Model C (later GP) to production. Without proper time to develop the machine, the company's resident genius Theo Brown had been pushed to release the tractor with three-row rather than four-row capabilities—a critical mistake that would later be rectified with one of the many improvements made during the GP's rather fraught years of production.[7] Despite some interesting innovations, the Model C/GP didn't seriously challenge the Farmall on the sales floor or in the field.

FARMALL F-30
This model is being unloaded in front of the Griffin Implement Company, an IH dealership, in a photograph dated November 11, 1936. *Wisconsin Historical Society / 25594*

J. I. Case and its resident engineers, however, gave the Farmall a run for its money in 1928. Its Case Model CC was a general-purpose tractor with less weight and more horsepower than the Farmall Regular. This caused considerable consternation at IH, and Oliver and Allis-Chalmers would add fuel to that fire with competitive models shortly after.[8]

The IH response was to increase horsepower across the board, and it had several machines in development around 1930. One was a 7,000-pound enlarged Farmall, the F-40. That model was ultimately scrapped, but management would put its smaller brother on the fast track to production in February 1931 with a $335,000 infusion of cash.[9]

The new Farmall F-30, which was a larger version of the Regular with a bigger 284-cubic-inch four-cylinder engine, would begin production at Farmall Works in October 1931.

The tractor featured a four-speed transmission, one more speed than the Farmall had. An optional high-speed gear set was offered in late 1934 (which is different than the overdrive offered for Cane variants in 1938). A low-speed attachment was offered for F-30s built on or after March 1937.

At 12 feet, 3 inches, the F-30 was nearly 2 feet longer than the Regular. It turned tightly, like the original Farmall, but took a 3-foot-larger radius to do so. Still, a turning radius of just over 17 feet was impressive for a tractor of that size. The McCormick-Deering 15-30 turning radius was nearly double that, at 30 feet.

The F-30 weighed about 5,300 pounds, which was 700 pounds less than the 15-30 and nearly 2,000 pounds more than the Regular. The F-30 engine used a 4¼x5-inch bore and stroke, making it a shorter stroke and a smaller bore than early McCormick-Deering 15-30s. The F-30 engine ran at 1,150 rpm. The additional 150 rpm was enough to give the F-30 engine more horsepower than the 15-30, and the extra weight combined with the slightly more powerful engine resulted in a bit more pulling power in the field.

The F-30 was originally proposed with a slightly less powerful engine. One of the first decisions on the F-30 concerned an increased-power engine and increased coolant capacity, and that decision received final approval on May 1, 1931. The decision bears a handwritten note stipulating that the serial numbers for the more powerful engines would be AA501 and up. Construction of the new engines was to begin at Tractor Works on July 15, 1931. Considering that only 623 F-30s were built in 1931, it's doubtful that any were made with the less powerful engine.[10]

The tractor was equipped with a power take-off and an oiled air filter, and it could burn distillate, kerosene, or gasoline. The manifold had heat ducts that routed exhaust gas to warm the Zenith carburetor. The carburetor needed to be warmed up to burn lower-grade fuels such as kerosene and distillate. Distillate was considered the most economical farm fuel by 1925. It didn't produce nearly as much power as gasoline, but it was much less expensive. Kerosene wasn't as economical as distillate, but some farmers still used it.

The F-30 engine's cooling system was comprised of an integral water jacket cast into the engine, a front-mounted radiator, a water pump, and a thermostat. The thermostat as

1932 FARMALL F-30 "DUCKBILL"

The early F-30s had a unique cover over the front of the steering gear, shown here on this early model.

Kinzenbaw Collection / Lee Klancher

1938 FARMALL F-30N

The F-30s were available in a narrow version, which had a 56.75-inch tread width.

Andrew Dawson Collection / Lee Klancher

FARMALL F-30 CANE WITH DUALS

This Farmall is working sugar cane fields operated by the United States Sugar Corporation in Clewiston, Florida, in April 1939. *Marion Post Wolcott, Library of Congress LC-USF34- 092054-D* well as canvas curtains or shutters regulated cooling. Early F-30s used an IH E4A magneto, while later examples had an F-4 magneto. A Bosch U-4 magneto was available as an option.

The model was available with a wide or narrow front. The rear tread width of the model had two sizes available: a standard or narrow-tread version.

The F-30 Cane Tractor variant was offered beginning in December 1934 and used various configurations of taller wheels and different front axles for higher-than-stock clearance. In 1938, a high-speed version of the F-30 Cane Tractor was offered, which added an overdrive gearing unit that increased the top speed to a blistering 16.75 miles per hour.[11]

An F-30 Rice Special was also built for use in Arkansas rice fields, and it had "Rice Special" stenciled on the left and right frame rails.[12]

PNEUMATIC TIRES

In 1933, the company authorized pneumatic tires as an option on IH tractors. Pneumatic tires were a solution to an increasingly-troublesome problem for the farmer. Paved roads were becoming more common in rural areas, and lugged steel wheels severely damaged pavement. Laws were passed prohibiting the use of lugged steel wheels on paved roads, and farmers were forced to use planks anytime they needed to cross a paved surface.

Rubber-coated steel bands known as overtires and easily detachable lugs were created in an attempt to solve this problem, but installing these devices was too time-consuming to be practical. Solid rubber tires were in use, mainly on industrial tractors, but they provided poor traction in wet conditions. Chains were a partial solution to this problem, but then the farmer was back to dealing with something that needed to be removed to cross pavement.

B. F. Goodrich experimented with rubber tires that were arched rather than air filled in 1931, but Firestone provided the solution in the form of the pneumatic—air-inflated rubber—tire. The Firestone tires were introduced on Allis-Chalmers tractors in 1932, and International demonstrated the tires on its machines as well.

The pneumatic tires provided good traction in all sorts of conditions and had the added advantage of an improved ride for the operator. In a 1934 trial of lugged steel and pneumatic tires, the University of Nebraska Tests found that steel had a slight edge in mowing alfalfa and grass, but pneumatic tires worked more acres per hour while plowing, cultivating, combining, and harvesting crops. Conversely, steel had a slight advantage in fuel economy.

Another advantage of the pneumatic tires was the ability to fill them with water to weight the rear end for traction. Rear wheel weights were commonly used, and water was a simple way to increase rear traction without lugging heavy weights about. In 1945, the Lift-All was adapted to the F-30 and offered as a retrofit kit.

The F-30 proved to be a rugged, maneuverable tractor, and it did well for both IH and the farmer. It was built and sold in respectable numbers until 1939 (with a few high-clears or specialty models built in 1940). The tractor slipped out of production when the Farmall tractor line was completely redesigned and the styled Letter Series was introduced.

RADIO-CONTROLLED F-30
International developed tractors that could be remotely operated using radio waves in the 1930s. This machine is operating at the Indiana State Fair in 1931. *Wisconsin Historical Society / 25495*

McCORMICK-DEERING W-30

Like the Farmall F-30, the McCormick-Deering W-30 was released in 1931 and was part of the company's effort to offer more power on its model line. The W-30 was an upgrade of the McCormick-Deering 10-20. The increased-power engine was bumped up to about 30 horsepower, hence the designation.

The W-30's engine put out a little more horsepower than the Farmall F-30 mill. The bore and stroke were the same at 4¼x5 inches, but the engine was rated for 31.63 horsepower at the belt, which was 1.3 horsepower more than the F-30. At the drawbar, the two were rated nearly equally. Both engines were governed to run at 1,150 rpm, although the W-30 was rated to run at 1,300 rpm late in production (which was achieved by using a new governor spring).

The W-30 engine used a combination manifold that allowed it to burn kerosene, distillate, or gasoline. A Zenith carburetor was used, as was an International E4A magneto. A thermostat-regulated radiator and water pump cooled the engine. Power was transmitted to the transmission via a 12-inch single-plate clutch, and the brakes were hand operated.

Although the engine was nearly identical to that of the F-30, the chassis was vastly different. The integral frame was used—meaning that the engine and transmission housings were part of the frame itself—but the front and rear axles and final drive were all unique to the wheeled 30s.

The transmission is a bit of an enigma, as early company records list a three-speed transmission. At some point in production, a four-speed transmission was adopted. In 1936,

SHOWROOM CONDITION W-30

The W-30 was a hot-rodded McCormick-Deering 10-20, part of IH tractor engineering's movement to increase power in its machines for the early 1930s. *Wisconsin Historical Society / 7227*

McCORMICK-DEERING I-30 AT WORLD'S FAIR
The Industrial line of tractors became a significant part of the IH income stream in the 1930s. *Wisconsin Historical Society / 49721*

a lower-geared four-speed transmission was fit, and serial numbers were changed with the improvement. The decision to make the change was made early in 1936, but a few of the older units were sold into 1937.

Options for the W-30 included pneumatic tires, high-altitude pistons, a lower low gear, differential lock, and a distillate-burning manifold.

Fewer than fifty W-30s were built in the first two years of production. The tractor was finally sold in quantity in 1934, when 2,634 rolled out of the factory doors. Sales remained steady until 1940, when the W-30 was being phased out in preparation for the W-6.

The W-30 served IH well. Although the company was able to use parts and technology designed for the McCormick-Deering 15-30, the cost of building a custom rear drive and front end made the W-30 expensive for the volume of sales produced.

McCORMICK-DEERING I-30

When the W-30 appeared, it was quickly pressed into duty on the Industrial tractor line as the I-30. It fit neatly in between the Model 20 and Model 30, giving greater flexibility to the line.

The experience the company had with industrial tractors led to a number of improvements to the I-30. The drawbar was strengthened and built to be more adaptable to mount equipment. Some I-30s were fitted with equipment to perform specialized tasks ranging from landing dirigibles to operating crosswalk plows. The I-30 was produced at Milwaukee Works.

The I-30's specifications closely resembled those of the W-30, which was the base model for the machine. The I-30 was slightly smaller than the Model 30 and weighed more than 2,500 pounds less with about 10 fewer horsepower.

The I-30 engine was an IH valve-in-head four-cylinder with a 4¼x5-inch bore and stroke. The engine was rated for 35 horsepower and governed to run at 1,150 rpm. Like most of the IH line, it used removable cylinder sleeves, ball-bearing crankshaft journals, and splash/pump lubrication. The cooling system used a radiator, water pump, and thermostat.

Options included a belt pulley, 575-rpm PTO, electric and gas lights, fenders, special wheel equipment, and custom-built equipment.

FARMALL F-20

The horsepower brigade continued with an upgraded version of the Regular known as the F-20. The F-20 had more power than the Regular, and a few more features, and the first of them were built on January 5, 1932.[13]

Introduced into the worst farm economy of the twentieth century, F-20 production ran from January to April 1932 then was suspended until December 1933. The machines built in December 1933 and January and February 1934 have unique serial number prefixes—TA rather than FA, which is on the others—and use Farmall Regular rear countershaft housings without cast foot platforms.[14]

As with other Farmalls, F-20s built on or after November 1, 1936, were painted red. The F-20 was upgraded with a more powerful engine and some additional features in September 1937.

The F-20's transmission used four speeds rather than the three of the Regular, and the engine used a bore of 3¾ inches and a stroke of 5 inches. The four-cylinder engine was governed to run at 1,200 rpm and used removable sleeves, ball bearings on the crankshaft, and a combination of splash and gear-driven pump lubrication. It also had an oiled air filter. The engine was started on gasoline and could burn either kerosene or distillate. It was cooled with a thermosyphon

FARMALL DUCKBILL F-20

The F-20 was an upgraded Regular and part of the company's efforts to increase power in its tractor lineup. Early Farmall F-20s had this distinctive front bolster cap. *Case IH*

system with a radiator. The clutch was an 11-inch single disk.

The F-20 came in four variants in addition to the standard model: a narrow-tread version, narrow-tread with wide front, regular tread with wide front, and special.

In June 1938, the first F-20 Cane Tractor was built. The tractor was a narrow-tread F-20 with the addition of a high-arch wide-tread front axle with new steering knuckles, steering knuckle arms, and drag link in the front as well as larger front and rear wheels and foot brakes. The F-20 Cane Tractor was manufactured for use with tall crops. The company decision lists Australia as a key market for this tractor. Interestingly, it could be equipped with a high-clearance attachment, which could be bolted to the front end. This attachment increased front axle clearance by more than 12 inches and widened the front axle tread nearly 6 inches.

1938 FARMALL F-20 WITH LIFT

This F-20 is equipped with a Farmer's Friend Lift. *Kinzenbaw Collection / Lee Klancher*

COTTON PICKER

International spent years developing cotton harvesting equipment. This is an experimental picker shown on the Hopson Plantation near Clarksdale, Mississippi, in 1937. *Wisconsin Historical Society / 24584*

"FARMERIZED" F-20

F-20s could be equipped with a non-adjustable wide front. This one is not stock and has most likely been modified by an owner. *Van Ruler Collection / Lee Klancher*

F-20 WIDE-FRONT TRACTOR

This is the Farmall wide-front tractor. There was also an F-20 Cane Tractor that was authorized to be built in 1938. It was based on a narrow tread model and featured a different front bolster and taller rear tires. According to company decisions, a high clearance version of this model was made available a bit later. *Case IH*

GOODYEAR TIRES

Rubber tires were added to F-20s beginning in 1932, with both Goodyear and Firestone tires available on their own wheels as well as wheels built by French & Hecht. *Goodyear*

T-20 CRAWLER

International added power to the TracTracTor model and changed the name from the Model 15 to the T-20. The little TracTracTor was small and maneuverable and was used in a wide variety of situations ranging from orchards to logging to grain harvesting. As you would expect, the crawler was more expensive than the wheeled tractors.

Priced around $1,500 in 1935, it was about $400 more than its wheeled counterpart, the F-20. Despite the high price, the T-20 was fairly popular. About fifteen thousand units were sold from 1932 to 1939, until the new line of streamlined crawlers choked off the old T Series.

The T-20's four-cylinder IH engine—which was the same as the one used in the F-20—had a bore of 3¾ inches and a stroke of 5 inches. The engine was rated at 31 horsepower and was governed to run at 1,250 rpm, 50 rpm higher than the F-20. An air cleaner, oil filter, and fuel filter were used, as was a three-speed transmission. The engine was started on gasoline and switched over to kerosene when warm.

The T-20 weighed 6,250 pounds and came equipped with a 543-rpm power take-off. The T-20 put out about 25 horsepower at the drawbar. Standard tread width was 41½ inches, although a 51-inch-wide tread version was available as well. A cab, canopy, belt pulley, front pull hook, and other options could be purchased for the T-20 on special order.

LOGGING EMPIRE CIRCA 1938

This T-20 is one of two of the crawlers used on the F. J. Jacks logging operation in northern Mississippi. *Wisconsin Historical Society / 59349*

T-20 WITH ORCHARD FENDERS

The T-20 TracTracTor is pulling a 300-gallon spray tank on John Crane's Fennville, Michigan, fruit farm in an image dated June 1938. *Wisconsin Historical Society / 59831*

DRY FARMER'S SON

This T-20 is at work in Oneida County, Idaho, in an image dated July 1940. *Russell Lee, Library of Congress LC-DIG-fsa-8a29271*

FARMALL F-12 (AND F-14)

Alexander Legge favored small farmers, and his influence helped drive the IH approach to its early twentieth century tractor development. The farmers of the day were struggling on two fronts: while confronted with the worst farm economy in history, they had to increase production.

The F-12 was designed to offer the latest advances in power farming at a much more economical price—$500 to $600 depending on the equipment.[15] The low price made the machine ideally suited to the times.

Preproduction machines were built beginning in August 1932, with regular production kicking off in January 1933. Serial numbers 533 and 600 to 608 had a preproduction version of IH's new 113-cubic-inch four-cylinder. The factory then switched to using an experimental Waukesha FL engine for the regular production models. Models with serial number 3035 and up were equipped with the production version of IH's 113-cubic-inch four-cylinder.[16]

The new F-12 engine would encounter a host of issues when burning poor-quality kerosene. Three out of four F-12s sold were kerosene models. Those that went to areas of the country with low-quality fuel would suffer catastrophic failures and premature wear. Most of the tractors sold in Little Rock and half that went to Memphis, for example, required rebuilds and new crankshafts. New manifolds and revised carburetors were added—one of the many production changes made during the life of the F-12.[17]

The engine was equipped with a new IH 1-inch downdraft carburetor, a Purolator oil filter, and an IH magneto. Approximately the first 2,000 F-12s built were equipped with an IH E4A magneto, and later examples were equipped with an F4 magneto.

1933 FARMALL F-12
Early production F-12s had Waukesha flathead engines. They are easily spotted as they are significantly shorter than the IH engine on later models. *Jack Bailey Collection / Lee Klancher*

Later references to the F-12 and F-14 indicate that both gasoline and kerosene versions of the tractor were made. It makes some sense that the dual-fuel manifold would be left off the machine to help keep the cost down. Use of the IH engine required extensive changes to the control levers, cooling system hookup, and exhaust and intake attachments. The main frame beams were changed as well. A total of thirty changes were listed on the specifications sheet, undoubtedly requiring a fair amount of retooling and retrofitting.

The Regular used a dropped-rear drive mechanism. On each wheel, an enclosed, oil-bathed bull gear was driven by the axle pinion. This added complexity to the tractor but increased ground clearance for cultivation. One of the drawbacks was that tread width was adjustable only by flipping the offset rear wheels.

SELECTING MACHINERY
The F-12 was designed to be a more affordable cultivating tractor ideal for small farms with modest budgets. Given the times, pretty much all farmers had modest budgets. This farm rehabilitation client is considering options at a dealership in Jackson, Ohio, in spring 1936. *Theodor Jung, Library of Congress LC-DIG-fsa-8a14181*

1934 FARMALL F-12
The later F-12s (built after June 1933) had an IH 113-cubic-inch overhead-valve four-cylinder engine. Three different cylinder heads were used during production. *Guy Fay Collection / Lee Klancher*

FARMALL F-12 WIDE FRONT

This F-12 is planting beans near Lake Okeechobee, Florida, in April 1939. *Marion Post Wolcott, Library of Congress LC-USF34-092066-D*

The engineering stroke of genius that eliminated the dropped rear drive was simple, not to mention the fact that it was someone else's idea. The Case Model CC and Oliver Hart-Parr Row Crop used larger rear wheels to maintain the ground clearance necessary for cultivating. For IH to compete, it needed to eliminate the dropped rear drive, and it did so with the simple solution of larger rear wheels.

This not only eliminated the dropped-rear drive mechanism, it also freed them from the restraints of a fixed tread width. Rather than fooling with clumsy reversible hubs, the rear axle of the F-12 was keyed so that the wheel could slide freely. The tread width adjusted between 44 and 78 inches. This eliminated the need to produce a narrow tread version of the tractor, which was an expensive way to meet the farmer's needs.

The Farmall F-12 was the first model to be used as a basis for both the wheel and cultivating tractors. From the F-12 forward, International wheel and cultivating tractors would become increasingly similar. By using one basic model, production costs were cut, engineering time was reduced, and retooling costs were slashed. By the 1950s, the limited differences between the standard tread and cultivating tractor became more subtle than in the past. Converting a row crop to a standard tread tractor required only a few parts to adapt the wide front end and lower the tractor a few inches.

The new concepts aside, the F-12's roots resided in early Farmall development. At one point, a lighter version of the Farmall was discussed. In May 1921, engineer Bert R. Benjamin wrote company executive Alexander Legge, suggesting that they build a tractor similar in size and power to the Fordson with a narrow front end. Benjamin's contention was that such a machine could plow as well as cultivate. At the time, it would cost the farmer at least $1,200 to purchase both a tractor to plow and a tractor to cultivate. Benjamin felt the new IH machine could be sold for about $700. He also felt that someone, probably Ford or International, would produce the tractor in the near future.

Benjamin's description was close to that of the F-12, a lighter tractor, weighing around 3,000 pounds, that sold for about $800 when introduced. These specifications closely mirrored those of the early Farmall designs in weight and power.

The company was enthusiastic about the new F-12 and felt it had more market potential than any of the previous Farmalls. The F-12 could be sold at a cheaper price than previous Farmalls, and Benjamin felt it was more profitable to operate than either animal power or the combination of a plowing tractor (typically a Fordson) and a cultivating tractor.

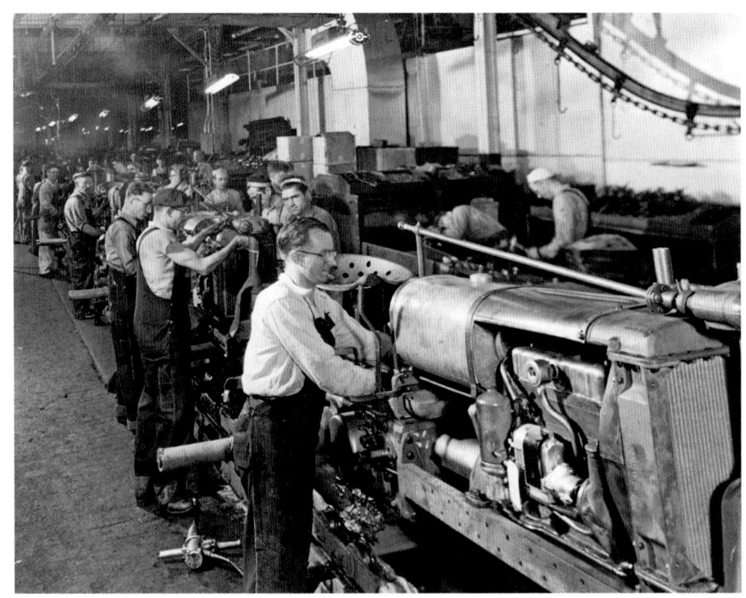

F-14 AT TRACTOR WORKS

This image dated June 24, 1938, shows F-14s being assembled. The first production F-14 was built on January 27, 1938, and the last F-14 was built on April 20, 1939. *Wisconsin Historical Society / 12234*

F-14

In 1938, IH decided to see if giving the F-12 a little more power would put more life into sales. The engine was governed at 1,650 rather than 1,400 rpm, and the gearing was adjusted to maintain the same vehicle speeds. Some additional improvements were made, including raising the seat and steering wheel to a higher and purportedly more comfortable position. Also, a new PTO pinion gear was installed.

In February 1938, it was determined that the modified F-12 would be known as the F-14. A belt pulley was added to the list of standard equipment, and the new F-14 was introduced in mid-1938. The decision notes that all F-12 accessories and implements would work on the F-14 except for the hydraulic lift attachment, which was a different unit for the F-14.

ELECTRICS

An electric starting and lighting package for the F-12 and F-14 was introduced in 1937 and modified in 1938. The modifications allowed the battery box to be located on either side of the tractor to accommodate implements. Two different part numbers were used for the kits for gas or kerosene models, but the sole difference between the kits was the choke and governor control brackets.

In 1939, the F-14 sold more than sixteen thousand units, a respectable number. The model had carved out yet another niche for IH tractors, proving that the market would bear a smaller general-purpose tractor.

1939 FARMALL F-14

The F-14 below is the upgraded version of the F-12. The major difference between the two was that the F-14 was rated to run at a higher rpm, giving a bit more horsepower. Note that the steering wheel shaft has been raised a bit and angles away from the hood. *Charles Klein Collection / Lee Klancher*

WORLD'S FAIR POSTERS

Colorful posters advertised the Chicago World's Fair in 1933 and 1934. *(Left) Library of Congress LC-DIG-ppmsca-80390 (Right) Wisconsin Historical Society / 39444*

RADIO CONTROL AT THE WORLD'S FAIR

When Chicago decided to extend the 1933 World's Fair into 1934, IH showcased this radio-controlled O-12 and an I-30 on this field west of Farm Machinery Hall. The operator sat on the deck of the little farmhouse. IH would write, "Will the farmer of the future be able to sit on his front porch while directing all his farm work? Will it be possible to sit in an office in Chicago or New York and direct the operation of fleets of tractors throughout the world?" *Wisconsin Historical Society / 6585*

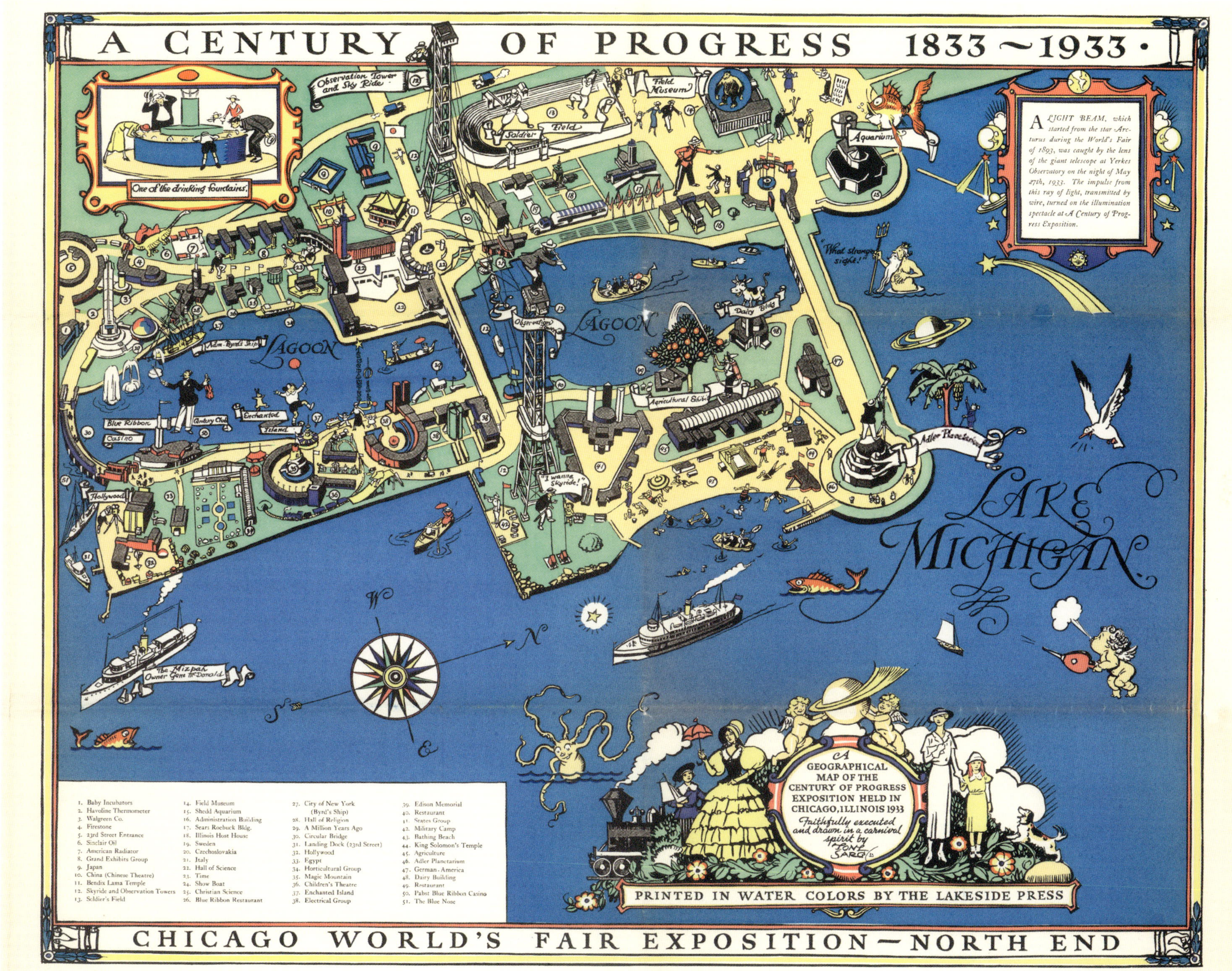

THE CHICAGO WORLD'S FAIR

The 1933 World's Fair Exposition celebrated Chicago's 100th anniversary and was extended to 1934. International Harvester had a large exhibition there. The show also featured popular celebrities and fantastic exhibits along the waterfront, with thousands of attractions such as a cigarette-smoking robot, the first MLB All-Star Game, a visit by the 776-foot-long Graf Zeppelin airship, and an exhibition of incubators with live human babies. *Creative Commons*

McCORMICK-DEERING W-12

The W-12 put out about 10 drawbar and 16 belt horsepower, similar numbers to the F-12. The W-12's standard equipment included steel wheels and slightly lower gearing than its siblings, the I-12 and O-12. Like the F-12, the W-12 was upgraded to the W-14 in 1938, with a higher rpm rating giving the tractor a bit more horsepower. This W-12 is pulling a disc harrow in Manitoba, Canada. *Case IH*

STANDARD TREAD 12 AND 14

From almost the first day IH began producing tractors, several different divisions and factories built an assortment of similar tractors with different engineering bases. The result was independently engineered and produced machines, requiring different tooling, independent development costs, and noninterchangeable parts. This drove up costs, and, in some cases, the tractor lines competed against each other.

With the standard tread and F-12s, IH could focus on building one basic layout, and easily modify it for both the cultivating and wheel tractors. The engine, transmission, and most of the large components in the 12 Series models were the same. The standardization of parts was hardly a new idea—a French general proposed it in the eighteenth century and Eli Whitney popularized the idea (though not the practice) when selling his guns to John Adams and Thomas Jefferson in 1801.

It was a practice that worked well for IH in the 1930s. The standard tread 12s did not sell as widely as the F-12s, but IH did not have as many developmental dollars to recover. Modest sales would return a profit, which was exactly why IH wanted to move to a single base engine and chassis design.

McCORMICK-DEERING FAIRWAY

The Fairway 12 was authorized in April of 1934, only a few months after the other 12s were approved. The Fairway 12 was simply an O-12 with steel wheels and foot-operated brakes. This Fairway 12 is at the Aurora Country Club with groundskeeper Ralph Teeter at the wheel. *Case IH*

The field, orchard, and industrial versions of the little 12s were authorized on the same day, December 4, 1933. They were introduced as 1934 models, and all three used the same engine as the F-12, with a slightly different frame.

These tractors weighed about 3,200 pounds, just like the F-12, and used hand-operated brakes and throttle, a three-speed transmission, and an 8-inch single-disk clutch. The frame was a two-piece unit bolted together in the middle. The front axle was a low-slung unit with tapered roller bearings for the wheels, and the rear axle was semifloating with diaphragm oil seals. Rubber tires were optional, and the cooling system was thermosyphon with shutters on the radiator controlling temperature. A swinging drawbar and pan seat completed the package.

The orchard version bore swoopy fenders and low-slung exhaust and intake pipes, while the industrial had different brakes and drawbar. Both wore rubber tires, while the I-12 had foot brakes and the O-12 brakes were hand-operated. The transmissions in the I-12 and O-12 had a tall third gear, which allowed the little tractor to sprint through the orchard or across the warehouse at just over 10 miles per hour. The

INTERNATIONAL I-12

The 12s were upgraded in 1937 to accept lights, making it likely that this machine was built after that time. This I-12 was photographed working at the Newark Airport in New Jersey. *Wisconsin Historical Society / 9475*

McCORMICK-DEERING W-12 AND I-14

The industrial version of the W-12 and the I-12 (later I-14) rode on rubber tires and used a taller third gear than the W-12.
Lee Klancher

McCORMICK-DEERING 1-12 WITH HOUGH SHOVEL

An I-12 at work with a Hough Shovel at the Farmall Works.
Case IH

W-12 rode on steel wheels, had a farm-type drawbar, and was geared a bit lower for field work.

The fourth addition to the standard tread 12s was the Fairway 12, which was simply an O-12 with steel wheels or rubber tires and an I-12 foot brake. The Fairway was designed for—you guessed it—use on golf courses. The Fairway 12 was authorized in April 1934.

In 1936, the 12s were upgraded. The 8-inch clutch was replaced with a 9-inch unit, an agricultural-type seat was fitted, foot brakes replaced the hand brake, and the wheels and tires were changed.

The front half of the frame was upgraded in 1937 to accommodate electric starting and lighting. The flywheel, steering gear housing, rear engine support, timing pin, and clutch pedal return spring were new, in addition to the new front main frame.

STANDARD TREAD 14

The 14s used the improved engine found on the F-14, with a variety of other component changes. The first W-14 and I-14 were built in March 1938, with the first O-14 not long after in April 1938. Note that the Fairway models were considered variants of the orchard model, so records are not clear when the first was built.

The little 12s and 14s provided International with an easily-convertible wheel tractor and made it easier for the company to offer such machines.

1938 McCORMICK-DEERING FAIRWAY 14

Tractors imported to New Zealand were often interesting mashups of models from the US, UK, and Australia. This Fairway 14 is a version of the orchard model modified for use on a golf course. *Heath Gander Collection / Lee Klancher*

McCORMICK-DEERING W-40

With the market's demand for horsepower steadily increasing, IH responded with a powerful wheel tractor, the W-40. The new tractor used a heavy-duty integral frame mated to the six-cylinder engine from the TA-40 TracTracTor. Seven experimental versions of this rugged machine were built for testing in 1934.

The W-40 was first designated the as WA-40, then the WK-40, and finally simply the W-40. Each variant is identified by a unique serial number code, and some bore badging designating "WA-40" or "WK-40." The model was built at Milwaukee Works.

The six-cylinder $3\frac{5}{8}$x$4\frac{1}{2}$-inch engine ran at 1,600 rpm and was rated for 45 horsepower at the belt and 28 horsepower on the drawbar. The engine was started on gasoline and switched over to kerosene or distillate. A 14-inch single-plate clutch transferred power to the rear end. The transmission was a three-speed, and the front and rear axles were similar to those used on the W-30. The tractor weighed in at 6,100 pounds.

McCORMICK-DEERING WK-40

The big 40s—the W-40, WD-40, I-40, and ID-40—used engines from International's crawler line in a subtly upgraded frame from the McCormick-Deering 15-30. *Murray Collection / Lee Klancher*

INTERNATIONAL IA-40

The industrial versions of the models with gas (shown) and diesel engines appeared in 1934. The I-40 and ID-40 were not produced in large numbers, with fewer than five hundred built. This IA-40 is pulling an Adams grader near Macon, Missouri. *Case IH*

WD-40: A NEW DIESEL ENGINE

By Jim Allen

Diesel engines were difficult to develop. They were hard to start, and making them work properly was an arduous task. Caterpillar and International Harvester broke through both diesel starting problems in the early 1930s, each following different roads to the same destination. Caterpillar beat IH to the diesel market by almost two years, introducing a diesel engine in the Caterpillar Diesel 60 in 1931. The tractor's four-cylinder, 1128-cubic-inch diesel engine was started with a small twin-cylinder gasoline-fueled engine. The small engines used to start diesels are commonly known as a "pony motors." They would both spin over and warm up the diesel.

IH opted for a dual-fuel engine that started and warmed up on gasoline, then could be switched to diesel. Both

1934 WD-40 PREPRODUCTION SERIAL NUMBER 503

This WD-40 is serial number 503, the third of three preproduction units hand assembled by IH engineers in 1934. These very early prototypes used a number of parts with unique part numbers as well as a radiator mount and water trap that differs from the production models. Skilled restorer Wendell Kelch restored the machine. The main case was split at the bottom, and the carburetor was so rare that Wendell and his team had to hand fabricate one using drawings and photos Eric had of the original experimental models. Interestingly, all three of the preproduction WD-40s survived and are in the hands of collectors. *Eric Best Collection / Lee Klancher*

PD-40 DIESEL ENGINE
The mid-1930s PD-40 had a 4¾x6½-inch bore and stroke, making 460.7 cubic inches and about 48 PTO horsepower. It had a very stout lower end architecture, with wet sleeves and five main bearings. *Jim Allen Collection*

methods resolved the starting issues. Caterpillar gets a nod, along with Standard Oil Company, for leading the charge to develop lubricants that would work with a diesel engine. Diesel fuel quality also improved in the 1930s, driven by the increasing market acceptance of diesel power. The diesel technology jump-start of World War II took care of most lingering problems.

THE IH DIESEL

Dr. Rudolph Diesel patented his version of the compression ignition process in 1895 and had legal teams all over the world protecting it. An expensive licensing agreement was required to use anything close to his system, and evolution of the principle was notably slowed as a result. The patent finally began expiring in 1908, and after Dr. Diesel's mysterious death in 1913, the world was largely free to use and refine what became known as the diesel cycle.

1935 WD-40 ENGINE
The yellow arrows indicate key elements on the IH diesel engine: (1) The timer control on the injection pump. (2) The starting lever, which is normally stowed until needed. (3) The starting mechanism shaft to which the starting lever is attached. *Jim Allen Collection*

McCORMICK-DEERING WD-40

The WD-40 was the first diesel tractor tested at the Nebraska Tractor Tests, and it produced 44 horsepower at the belt. *Murray Rowlands Collection / Lee Klancher*

WD-40 ENGINE

The International PD-40 diesel engine used a unique starting system that ran on gasoline before switching to diesel fuel. *Murray Rowlands Collection / Lee Klancher*

The Gas Power Engineering Department of IH began experimenting with diesels in 1916 and by 1919 had built a 10-horsepower, single-cylinder prechamber Indirect Injected (IDI) diesel. It didn't lead anywhere directly, but research continued, and in 1927 in Germany—the spiritual home of diesel technology—IH engineers purchased a Dorner four-cylinder diesel for testing. The Dorner taught IH engineers enough to build three similar experimental test engines, the smallest of which was installed in a tractor and tested for two years at the Phoenix, Arizona, proving grounds. Along the way, more German engines were purchased for tests, but by 1930 IH had a unique design that was as good as, or better than, the German diesels. In the process, they also designed a bypass-controlled fuel injection system.

As IH saw it, the two base problems were sufficiently downsizing a diesel for use in tractors and solving the starting issue. The downsize was done by the end of 1930. For the starting problems, IH developed a system that combined a gas engine with a diesel, and it was complete by the end of

1932. The engine started on gasoline and then switched over to diesel after it was warmed up.

EARLY PRODUCTION: THE PD-40 AND PD-80

The first IH diesels were part of a new engine family. Diesel, gasoline, and distillate-fueled engines shared new heavy-duty architecture, so the main components of all could be manufactured on the same tooling. The first production IH diesel was built in April 1933 and the bare engine was designated PD-40. In this era, the engine designation changed to reflect the installation, so a PD-40 in a diesel crawler would be called TD-40. In a wheeled tractor, it was the WD-40. A diesel power unit was UD-40, and an Industrial diesel tractor was ID-40.

The big innovation, of course, was the gas start, and the nearby illustrations provide a clearer picture of how it worked. The main feature was the auxiliary combustion chamber connected to the diesel chamber by a starting valve. With the starting valve open, the added volume of the auxiliary chamber dropped the compression ratio from 15:1 to about 5:1. The auxiliary chamber had a spark plug and its own ignition system. The gasoline intake tract was connected to the diesel system via an air valve and a tiny carburetor supplied enough air and gasoline to run the engine at about 400 rpm. The starting valve lever also enabled or disabled the magneto.

The first IH diesels were manually started and had no electric start. Many gas tractors of the era were crank started, and it was not significantly more difficult to start an IH diesel in gas mode compared to a gas tractor of equivalent displacement. The injection pump had a timer that counted the number of engine revolutions (about 700), and it could be used optionally to automatically trigger the system to switch to diesel operation after approximately two minutes running on gasoline. In most cases, two minutes of running would warm the engine enough to run on diesel.

A six-cylinder diesel, essentially the PD-40 with two cylinders added, began development in 1933 and debuted in July 1936. It was a stout, seven main bearing, 671-cubic-inch engine (4¾x6½-inch bore and stroke) with wet sleeves. It had a lot of design evolution compared to the PD-40,

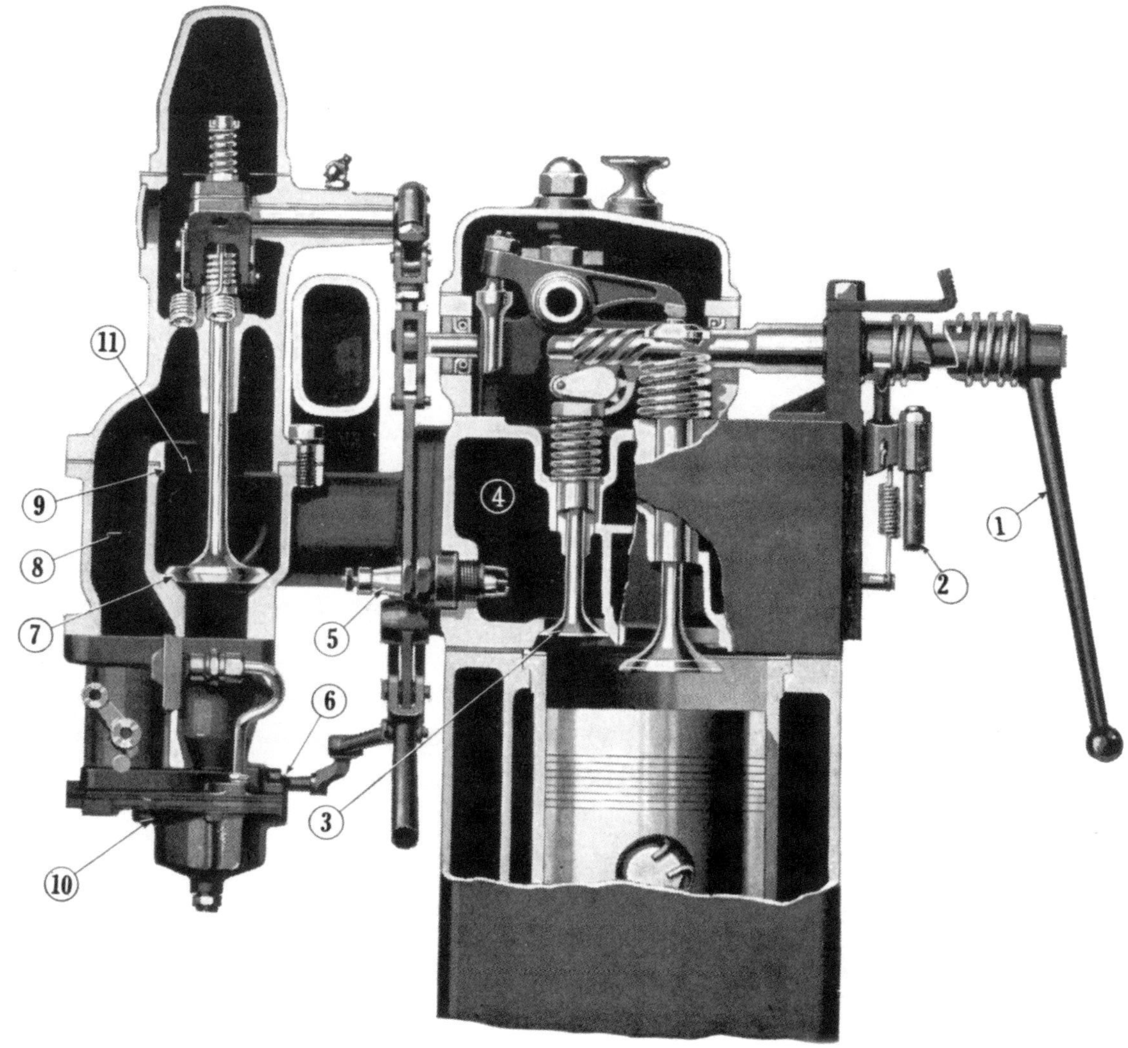

PD-40 INNER WORKINGS

A cutaway shows how the system worked on the PD-40. (1) A spring-loaded lever opens the starting valve (3) to connect the gas and diesel combustion chambers and move the gasoline system poppet valve from its lower seat (7) to its upper seat (9). It opens the gasoline valve (6) and allows fuel to gravity feed into the carburetor (10) and permits airflow through the intake passage (8) and into the combustion chamber. The release rod (2) operates a valve on top of the injection pump (not shown) that shuts off diesel fuel flow. *Jim Allen Collection*

most notably an improved combustion chamber. The PD-80 cranked out a maximum of 100 horsepower, 80 horsepower continuous at 1,400 rpm. It didn't immediately find a home in tractors but was offered as a power unit starting in 1937. A short-lived PD-35 four-cylinder variant appeared in the 1937–39 TD-35 "budget" crawler. The bore was ¼ inch smaller than the PD-40, but it made only 4 horsepower less.

INTERNATIONAL T-40 AND TD-40 CRAWLERS

While small crawlers were great for jobs around the farm, the construction and logging markets had a need for large, powerful crawlers. For IH to be player in these areas, it needed a big machine, and the T-40 and TD-40 TracTracTors were the entrants. They were developed early in 1931, and the T-40 used a six-cylinder engine. The TD-40 used a four-cylinder diesel engine in the same chassis. Both were built at Tractor Works.

In 1933, the gasoline version became known as the TA-40. The diesel TD-40 was introduced. The six-cylinder engines built in 1937 and after were equipped to burn kerosene or distillate. Note that advertisements sometimes listed the tractor simply as the T-40.

The TA-40 and TD-40 were physically larger and much more powerful than the T-20, with the gas tractor rated at 51 horsepower, with 44 horsepower available at the drawbar. The tractor was more than 2 tons heavier than the T-20, with the gas model weighing 11,200 pounds and the diesel an even 12,000 pounds.

A five-speed transmission was used, allowing the operator to exactly match power to load. Top speed was a galloping 4 miles per hour. The six-cylinder engine's bore was 3⅝ inches and the stroke was 4½ inches. The engine was governed to run at 1,600 rpm. A radiator and fan regulated by a thermostat cooled the big six.

The tractor was available in a standard and wide-tread version, and sold for nearly $3,000, making it an expensive vehicle at the time, although not out of line for large crawlers.

INTERNATIONAL T-40

An IH crawler equipped with Bucyrus-Erie bull grader at work in the Sahara Mine in Carrier Mills township, Illinois, in 1937. *Wisconsin Historical Society / 7272*

The T-40 and TD-40 could be equipped with an assortment of Bucyrus-Erie blades, transforming the crawlers into bulldozers. Options included a power take-off, enclosed cab, electric lights, various sizes of track shoes, and a canopy top.

INTERNATIONAL T-35 AND TD-35 CRAWLERS

The TracTracTor T-35 and TD-35 were slightly less powerful versions of the big 40s. They were first built in December 1936 and continued production until the styled TracTracTors appeared in 1939.

The T-35 used a six-cylinder engine that could be equipped to burn distillate, kerosene, or gasoline. Burning gas, the big six put out 37 horsepower in Test G at Nebraska, while the TA-40 put out 44 horsepower in the same test, also burning gas. The six-cylinder had a 3⅝x4½-inch bore and stroke and ran at 1,750 rpm.

The TD-35 used a diesel engine that produced 42 horsepower on the belt at 1,100 rpm. At the drawbar, the TD-35 put out about 37 horsepower. The TD-35 put out about 9 fewer horsepower than the big TD-40 on the belt and weighed a few thousand pounds less as well.

INTERNATIONAL T-35 DIESEL

The T-35 was introduced in 1936 and was simply a lower-powered version of the T-40, available with either gasoline or diesel engines. This one is shown cultivating potato fields in the San Joaquin Valley in California in February 1939. *Dorothea Lange, Library of Congress LC-USF34-019022-C*

McCORMICK-DEERING
FARMALL

THE MECHANICS
Marion Post Wolcott, Library of Congress LC-USF34-053716-D

CHAPTER FIVE

EMERGING FROM THE STORM

THE LETTER SERIES 1939–1942

"The first year I was married I had a good crop, one kid, and five dollars. Today I have nine kids and not even a job. It'll be a miracle if I don't starve."
—John Leard, June 19, 1939, interview with Frank Massimino, WPA

The farm recovery from the travails of the Great Depression was a long, slow process. Many never did recover, beat down by lean years, barren lands, and emptied bank accounts.

South Dakota farmer Joseph D. Lacher was one of the gritty few who survived. To do so, he endured total crop failures as well as skies blackened by dust and hordes of grasshoppers. Dust stalled his truck and buried his equipment and barn door in 1934. When the soil was so poor it would only produce thistles, he harvested them to feed his cattle. The low-quality fare gave the herd "awful diarrhea."[1]

In February 1936, negative 35-degree temperatures killed his truck engine while he was headed to town to get coal, leaving him with a very cold walk home. By summer, the drought was back in force and on July 9, 1936, the thermometer showed 115 degrees, scorching his crops.

HARD TIMES

While times on the farm improved in the late 1930s, they were still hard. The farmers that survived often had aging equipment, like this converted Lincoln car used on a farm in Ouray, Colorado, and shown in September 1940. With the farm outlook improving, these people were ripe for a new machine. *Russell Lee, Library of Congress LC-USF34-037652-D*

Heavy snows of 1937 made life on the Lacher farm difficult, but it gave them a good leg up on moisture and made their harvest passable. In 1938, the family added one hundred sheep, cleared the ground of invading thistles, and planted corn in its place. Grasshopper bait worked to keep the pests away, but a lack of rain meant his barley and oats were of minimal quality, the wheat harvest was only for seed, and the corn was a total loss.

A decade of such trials and tribulations meant many farmers gave up on their barren lands and went elsewhere. Lacher "kept a level head and planted on because he understood that the drought would break and prosperous times would return."[2]

The leaders at International Harvester shared Lacher's optimism, at least in the gussied-up accounts in company magazines and annual reports. Sales-oriented CEO, Sydney G. McAllister, had replaced the wise old cowboy, Alexander Legge, in 1935. He stepped into hard times but was gifted with a developmental program for a new line of tractors. Even as late as 1938, however, he had to work to offer hope in his official missives. "The company began 1938 with high hopes," he wrote in the last 1938 issue of *Harvester World* magazine, "but

A NEW MODEL EMERGES

In September 1941, this group of leaders from the Two River Non-Stock Cooperative, an FSA co-op located near Waterloo, Nebraska, had a new Farmall M demonstrated for them. *Marion Post Wolcott, Library of Congress LC-USF34-059481-D*

TIMELESS MACHINES

The styled Farmalls offered the farmer of the day an irresistible blend of the latest features, implements built specifically for the models, and eye-pleasing styling. This one is pulling a No. 64 combine circa 1950. *Wisconsin Historical Society / 7803* the rapidity of the decline was so great that the management was compelled to make drastic curtailments of production."

Farmers found plenty of challenges out in their fields. The dust storms and insect hordes continued in 1939, and wide swathes of the country remained in a period of drought. Lacher planted new crops and reported good yields of barley, oats, and wheat. Despite his struggles, 1939 was the first successful year of the decade.

The year was a big one for International Harvester, at least from a machine perspective. It introduced the Letter Series, arguably its signature line of machines. The tractors sported innovative features combined with a functional, streamlined look. The gangly appearance of the earlier Farmalls was replaced with the graceful design created with the help of Raymond Loewy, one of the era's most prominent industrial designers. The new line was part of a company-wide design move that included a new logo. The machines were produced for more than a decade, and the basic form endured until 1958, when the curvaceous lines of International tractors were replaced with the squared-off look that lasted to modern times. The logo survived, with minor modifcations, until 1985, when the IH tractor division was merged with J. I. Case.

FAMILY PLANNING

Multiple programs offered financial assistance and advice for farmers. The Dixon family, shown in November 1939, switched from wheat farming to raising livestock at the advice of one of those programs.
Arthur Rothstein, Library of Congress LC-USF34-029146-D

The line also marked a significant achievement in International's decade-long effort to serve the world's vast market of small-acreage farmers—an initiative long championed by Legge. The F-12 and F-14 were the first steps in the right direction but proved to be not quite small enough. The all-new Farmall Model A, Model B, and a small combine brought cutting-edge technology to America's 3.2 million farms of less than 70 acres.[3] The farmer with larger acreage had plenty of exciting new machinery to consider with the H and M, as well as redesigned crawlers. The entire line was complemented with implements designed specifically for the machines.

The fanfare of the new model launch was dimmed slightly by the fact that IH had a less-than-appealing sales year, with the blame for that pinned on the need to liquidate the old inventory to make way for the sleek new designs.

Better times were ahead, but you wouldn't have known it on the Lacher farm in 1940. More dust storms and grasshoppers wreaked havoc with the family's crop, and a hailstorm rendered what was left a total loss. "A year's work gone in five minutes," Lacher wrote in his diary in 1940.[4]

1940 was unfortunate for Lacher, but other farmers fared better, with American farm income up 8.4 percent over the previous year. The improving conditions were enough to encourage farmers to upgrade aging machinery, and IH reaped the benefits. Its profit from tractors soared from $48.9 million in 1939 to $81.2 million in 1940.[5]

The following year would brighten prospects even further for farmers. While the world was enmeshed in the challenges and strife of World War II, crop prices and soil conditions were improving. Some of that was thanks to Mother Nature gifting more rain and less extreme weather, while the rest

A TRACTOR FOR THE MASSES
The Farmall A was intended for the millions of American small farms that did not have tractors in the early 1940s. This one is at work in Copiah County in the 1940s. *Luther Hamilton Sr. / Mississippi Department of Archives and History*

was due to farmers learning how to protect their soil. Lacher planted a mix of crested wheat, brome grass, and oats, and the roots helped anchor the soil to the land. Farmers in Texas went to more extreme measures, breaking hard pans of sand with chisel plows and planting drought-defying African grasses.

Wartime crop demand would finally lift the farmer to better years, as increased yields and income deepened their pockets. As always, improvement for the farmer was the key to success for agricultural equipment manufacturers. While IH found great demand for its products in the 1940s, shortages of raw materials and government rationing related to World War II would make selling machinery a tremendous challenge.

HIGH-PRICED OLD IRON
During the height of World War II, tractors were carefully rationed and buying a new one required a permit. Older machines like this Farmall sold for exorbitant prices. Auctioneer Joe Moore is shown at work near Downs, Illinois, in a photograph dated November 11, 1946. *Frank Bill / McLean County Museum of History*

THE TIMELESS (AND MILDLY MYSTERIOUS) CONTRIBUTIONS OF RAYMOND LOEWY

As a young boy, Raymond Loewy filled his school notebooks with sketches of airplanes, cars, and locomotives. Born in Paris in 1893, he grew up enamored with the new machines emerging into society. His love of machinery resulted in teen hijinks, such as sneaking rides on the first buses running in Paris to experience their 20-mile-per-hour top speed firsthand.[6] He also took to riding bicycles—which at that time were high-wheeled, fast, and dangerous.

His interests in vehicles prompted his parents to enroll him in engineering school. Given his lifelong obsession with appearance over form (a trait critics seized upon later in his career), it may have not been a good fit. This potential issue never came to a head, as his college education was interrupted by a call to serve. He was sent into the first World War as a private in the French army, served honorably, and came home as a decorated captain.

When his parents were killed by the Great Influenza epidemic of 1918, Loewy decided to follow his older brothers to America and booked passage to New York with only forty dollars to his name. While on board, he auctioned off a sketch he made during the journey, and the purchaser put him in touch with the editors of *Conde Nast* magazine.[7]

The introductions led Loewy to land work as an illustrator for *Vogue* and *Harper's Bazaar.*[8] His Art Deco illustration became quite well-known in New York fashion circles, and his contacts in the magazine world led him to a short stint designing window displays for Macy's. He later said that his transformation of Macy's cluttered front windows to more spare arrangements was met with such strong disapproval that he quit before he could be fired.[9]

While doing magazine work in 1927, Loewy met Sigmund

SS1 LOCOMOTIVE

Loewy designed several locomotives for Pennsylvania Railroad (PRR), including the GG-1 and the steam-powered SS1, shown here in 1942. He also designed the bar lounge for the PRR Broadway Limited. *Steve Harlow*

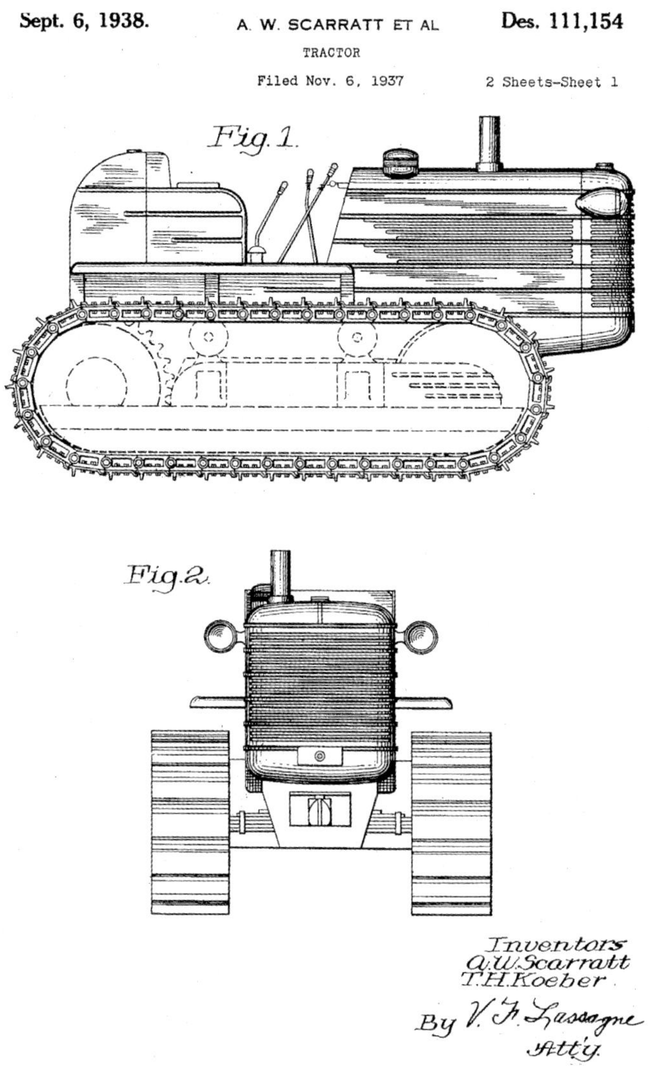

IH CRAWLER STYLING

Loewy's design contract with IH stipulated that the company held the rights to his creations. IH internal designers T. H. Koeber and A. W. Scarratt filed the patent for this Loewy styling of a crawler in November 1937. *US Patent*

Gestetner, whose company built duplicating machines that, according to Loewy, were ugly and smelled bad. Loewy offered to redesign them for $500. When Gestetner accepted, Loewy plopped $100 worth of modeling clay on a tarp in his living room and designed a sleek enclosure for the machines.

He was struck that there was opportunity for an enterprising sort to build a career transforming the rapidly growing number of new (and typically ugly) mechanical devices into more eye-pleasing forms.

RAYMOND LOEWY AND SKETCH

With a passion for powered machines that dated back to childhood, Loewy was involved in creating a number of influential vehicles. This photo from 1948 was made for a *Time* magazine cover story. *Bernard Hoffman / The LIFE Picture Collection / Shutterstock*

This commercial art form would become known as industrial design. As the business world came to understand that products that look and feel better, sell better, they turned to Loewy and his contemporaries like Henry Dreyfuss, Norman Bel Geddes, and Walter Dorwin Teague for designs. This is not to say that sales came easy. Loewy later bemoaned the fact that he spent several years on the road plying his wares. "I remember long gloomy winter nights in Cicero, Illinois, feeling sick, tired, and discouraged trying unsuccessfully to sell industrial design to minor but tough Midwestern manufacturers."[10]

By 1933, Loewy had assembled a reasonable list of industrial design clients that included Gestetner, Hupp Motor Car, and the appliance division of Sears. His big break came with his redesign of the Sears Coldspot refrigerator. The machine was a runaway success. Sales increased five-fold. In a 1949 cover story about Loewy, *Time* magazine would dub the design "a single smooth, gleaming unit of functional simplicity." Loewy and Sears decided to update the look of the refrigerator every few years, which provided another regular boost in sales.

AUTOMOTIVE CZAR

When Loewy worked for Studebaker, his office was housed within the factory and executives had to make appointments to be admitted. *Hagley Museum & Library*[ii]

Loewy's love of vehicles lived on, and he filed a patent for his design of an automotive hood in 1928. After landing a job drawing a trash can for Pennsylvania Railroad (PRR), he convinced them to take on more ambitious projects.[11] He drew PRR's stunning S1 streamlined locomotive as well as the gorgeous interior of the Broadway Limited railroad car.

He also had an encounter designing cars for the Hupp Motor Company, most famously the Hupmobile. He would later complain that his encounters with car engineers consisted mainly of them telling him what he couldn't build. Loewy's designs for Hupp were stylish but not what he had hoped. They were also slow selling, disappointing Hupp. The company and the designer parted ways; Hupp quickly declined and was forced to stop car production in 1937.

When Studebaker approached Loewy in 1938, he would agree to do the design work only if the Loewy design office was housed in the Studebaker facility. Studebaker executives were allowed to visit by appointment only. The arrangement would give Loewy a greater amount of control over the cars. Trailing the larger automakers in the industry by a significant margin and eager to find a way to gain market share, Studebaker agreed.

Such arrangements seem extreme, but history would show relationships like this could work well. The best example in the agricultural industry is John Deere and Henry Dreyfuss. The relationship between Dreyfuss and Deere CEO William Hewitt was so close that engineers required Dreyfuss to approve the final build before starting the line. That relationship worked, and the fact that Loewy required such a large amount of input at Studebaker is indicative of the viability of such partnerships.

About the time that Studebaker approached Loewy, he was also being courted by International Harvester. The IH team knew its new line of tractors needed to be revamped, and in-house efforts to style the machines were entirely unsatisfactory. At some point in the late 1930s, Loewy was asked to pencil a new look.

The fascinating thing is that little documentation has survived. The documents in the McCormick-International Harvester Collection held at the Wisconsin Historical Society have very few references to Loewy. A number of other institutions have Loewy archives, but those also have only scraps.

1937 STYLED IH TRUCK

Loewy's career with IH began in the truck division, with his work appearing on the 1937 line. In June 1936, he started working on tractors and power units. Due to his work with Sears, his IH contract stated he would not work on IH refrigerators. *Wisconsin Historical Society / 142718*

In several books written by and about Loewy, the accounts of the Letter Series redesign are very thin and occasionally technically incorrect. A variety of people have sent inquiries looking for more information.[12]

Loewy stated that the connection between he and IH began with his friend, Fowler McCormick, who pulled him into the company to do work.[13] One could easily imagine the two worldly, educated, and famous men moving in similar circles.

Loewy's first work for the company was designing a motor truck and was done prior to June 1936. He was contracted to act as the engineering stylist and designer for farm implements, tractors, and power units as well as trucks by IH's Edward A. Johnston on June 25, 1936.[14]

Prototypes of Farmall tractors with styled sheet metal exist in the archives beginning in spring 1937, and it is sensible to conclude that these machines were drawn by Loewy. The photographs of the prototypes from that time clearly show the transformation from the exposed mechanicals of the F-Series machines to a more finished design.

The earliest efforts are a bit awkward, and the new designs weren't met with universal acclaim. By spring 1937, Johnston complained that Loewy's designs "would be expensive [to build]."[15]

More clues to the evolution are found in the patents from that time. A patent for the crawler sheet metal was filed on November 6, 1937, by A. W. Scarratt and Theodore H. Koeber, who were members of the IH design team. A later patent for the hood lists Raymond Loewy as the primary patent holder, with T. H. Koeber as the secondary name. Note that IH's contract with Loewy stipulated that IH owned the rights to the designs he produced, so there was no need for IH to include Loewy on the filing.[16]

In an internal memo from Raymond Loewy/William Smith Inc. dated January 9, 1967, Fred Toerge reported about

RAYMOND AND HIS DOG

In the 1930s, Loewy spent most of his time in his office in New York, but he also kept a home in France, about an hour's drive south of Paris, at a place he called La Cense, which is now a horse training center.
Hagley Museum & Library[iii]

a trip to IH in which he met with Vice President R. H. Burnside and Director of Design T. H. Koeber.

"Had a good meeting with Mr. Burnside. He has pleasant memories of RL / WS . . . just as the conversation was getting interesting Koeber came on the scene."[17] Note that according to Toerge's itinerary, T. H. Koeber was not scheduled to be in that meeting.[18]

"Koeber has been with International Harvester for 30 years and now heads up their design group. I got the impression he is very hostile to consultant firms. According to Koeber, he last went to an outside source (Chapman) eight years ago. He said he would invite us to bid on any projects that require outside consultants in the future."[19]

When Loewy came on the IH design scene in 1936, T. H. Koeber was new to the company. The design of the original Farmall tractor was exposed and mechanical. The response we know about from the engineering team was that Loewy's improvements were expensive. The internal design team promptly took his suggestions, filed patents, and made sure their names were attached.

While not a personally flamboyant man—Loewy was soft-spoken and genteel—he was bright, opinionated, and not afraid to speak his mind. When asked by a manufacturer to draw sketches of a new television, he took the job and quipped, "Fine. I spent $2,000 on my own set and it hasn't worked right since I bought it."[20]

His tart opinions were delivered about everything from ocean liners and cars to charcoal grills and dish soap bottles. He also had opinions about the state of tractor design. "If you want me to do a big thing like a tractor—there are so many obvious things you could do to make it better-looking that I would take it for very little."[21] He made that statement in 1949 and was likely being dramatic—something he had a penchant for doing.

His 1936 contract with IH was hardly cheap, with a total cost of $50,000 cited on the 1936 five-year agreement. When adjusted for inflation to 2023 dollars, that works out to a cool $1.1 million.

It is possible that the combination of a personal connection with a rising executive and descendant of the company founder (Fowler McCormick), a pay rate far above and beyond that of a mere staff member, and the likelihood that his comments about the state of tractor designer were acerbic, one could easily surmise that Loewy quickly became at odds with the IH internal design staff.

Note that T. H. Koeber drew sketches of variations on the original Letter Series tractors in the 1940s or 1950s, likely during the same decade that he met Toerge. The sketches are variations on the original Loewy design, with some of the prevalent touches of the day incorporated into the original. John Deere engineer and artist Theo Brown sketched similar interpretations of the Henry Dreyfuss work on the Model A and B. Designers riff of each other and are naturally drawn to transforming and shaping the work of others, particularly when it takes place on their turf.

Gregg Montgomery was an IH industrial designer from more modern times. He came to IH in the 1970s and worked with IH industrial designer Larry Shinoda. When times got tough in the 1980s, he moved to doing contract work for IH and founded his own design firm, Montgomery Design International. He was also struck by the lack of information left about Loewy's contributions. "Everyone knew Raymond had worked for IH," Greg said. "Everyone talked about the different things he worked on. If you look into the history—and even his history—you don't find a lot about his work on tractors and trucks."[22]

Montgomery decided to write Loewy a letter, explaining his position in the company and asking Loewy to provide more details on what he did for IH. Loewy's response was a brief letter, in which he noted he was close friends with Fowler McCormick, who had introduced him to the company. He added that he drew the IH logo on a menu while in the diner of the Broadway Limited—a train, he noted, he had also designed.

On the letter to Montgomery, Loewy sketched a little logo that was different from the IH logo on the letter to Montgomery, and asked if anyone had caught on that he intended it to look like a man on a tractor.[23]

His last paragraph stressed the importance of spontaneous human inspiration when creating trademarks and warned of the dangers in the use of computers in logo design. Given the rise of artificial intelligence (AI) in design in present times, his words ring with prescience. He also promised to follow up and send Montgomery more details. He never did.

Loewy would later redesign IH dealerships as well as the IH logo, and there is more to come in later chapters on the fast-talking French gearhead and designer who helped IH shape its image. Despite the lack of sketches and other ephemera from the time, the Letter Series tractors that resulted from Loewy's work with IH in the 1930s left a cast iron legacy as some of the most important machine designs of the twentieth century.

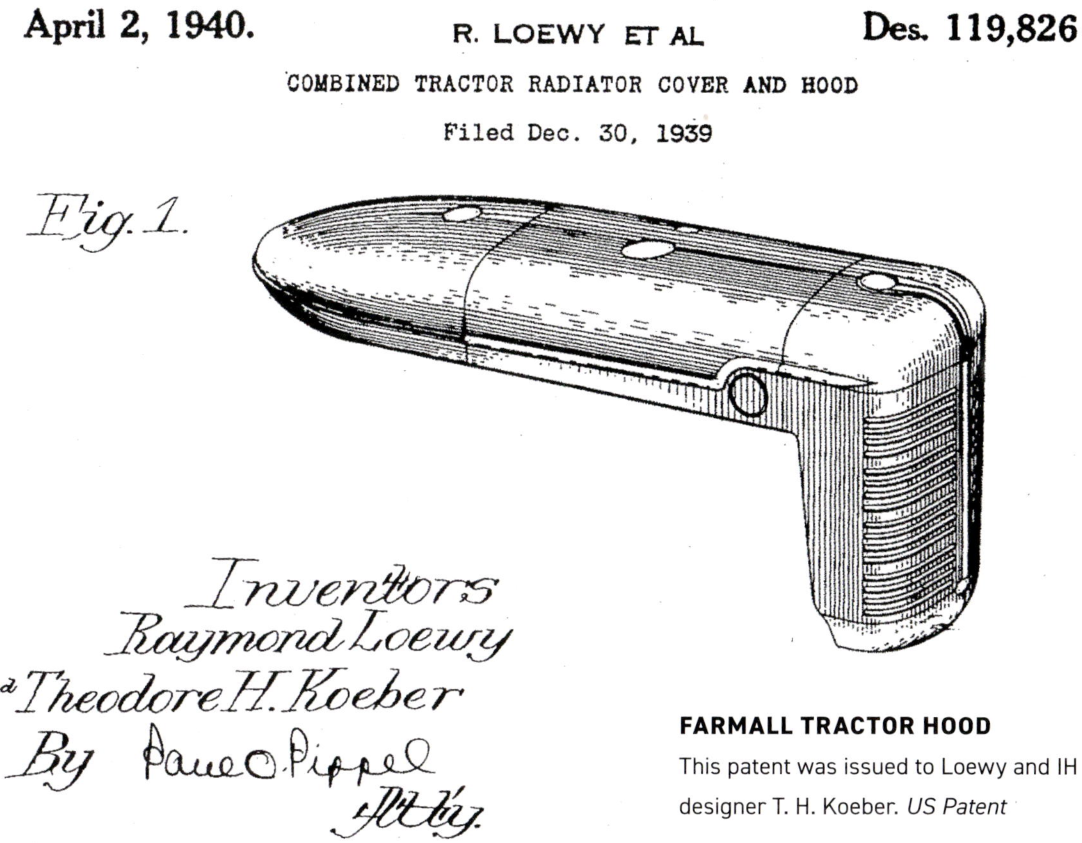

FARMALL TRACTOR HOOD

This patent was issued to Loewy and IH designer T. H. Koeber. *US Patent*

F-21 PROTOTYPE

In March 1936, the basic design of the new IH model combined the adjustable axle from the F-12 and a new lift. The serial numbering on the photograph indicates nine were produced. *Wisconsin Historical Society / 125528*

F-22 PROTOTYPE MAY 1937

Styling made its way to the new IH model in 1937. Experimental model number 3538 weighed 3,427 pounds. Derided as expensive by sales executive John McCaffrey,[iv] production of this model was nevertheless seriously debated. The tractor was a significant upgrade from the F-20, with more horsepower, improved brakes, and a wider range transmission as well as adjustable tread width, pneumatic tire compatibility, hydraulic lift, and more. Development would continue for several years—to the model's benefit. *Wisconsin Historical Society / 12204*

F-22 PROTOTYPE JULY 1937

By July, the F-22 styling was a bit cleaner. This is experimental number 3580, which weighed in at 4,520 pounds. *Wisconsin Historical Society / 12201*

LETTER SERIES DESIGN EVOLUTION

Beginning in the mid-1930s, IH started to develop its next new machine. The original Farmall was two decades old, and an update was due. Early experimentation was pretty simple: the engineering team grafted the straight axle from the F-12 to the F-20 chassis and built a machine called the "Lightweight F-20." This morphed into the F-21 in 1936.

The next variant, the F-22, was a more radical leap forward, with a frame design similar to that of the production Farmall H and M as well as sheet metal engine coverings that gave it a streamlined (if not entirely handsome) look. Experimentation on this front continued into 1938, with the F-15, 2-F, and 3-F experimental models built and tested.

IH had experimented with sheet metal for some time, notably with the International 8-16, but it was Oliver that brought the look to row-crop machines with its 1935 Model 70. The machine featured louvered metal covers over the engine grafted to a gracefully tapered front grill that gave the machine style and elegance.

Taking this a step further in 1938, John Deere hired famed industrial designer Henry Dreyfuss to redesign its new Model A and B tractors. The elegant front end of these iconic models arguably became a signature feature of the brand, and the relationship between Deere and Dreyfuss would remain tightly knit until the early 1970s when Dreyfuss died.

As the Farmall M and H prototypes began circulating, IH leadership took efforts to streamline and style its new machines. In the mid-1930s, IH decided to follow the practice of railroads, cigarette companies, and others and hire a high-profile industrial designer. They turned to Raymond Loewy, a brilliant, independent, and infamously sales-oriented stylist and *bon vivant*. One of the pioneers of the nascent industrial design movement, Loewy blended design talent with an unrivaled penchant for self-promotion. Worldly,

F-22 PROTOTYPE FEBRUARY 1938

Number 3722 was equipped with fenders and lights and weighed 3,820 pounds. *Wisconsin Historical Society / 12196*

2F PROTOTYPE AUGUST 1938

The F-22 became the 2F, of which several variations were built. This 2F is quite light, at 2,700 pounds, and is number 3800. *Wisconsin Historical Society / 12189*

3F PROTOTYPE NOVEMBER 1938

The prototypes for the Model M were known as the F-32 and later the 3F. This example is number 3830, photographed November 17, 1938. *Wisconsin Historical Society / 125812*

2F WOOD MOCK-UP

This December 9, 1938, image shows a wooden mock-up of the new Farmall. *Wisconsin Historical Society / 12193*

well-dressed, and quick-witted, Loewy was the perfect match for the pomp and circumstance that prevailed in IH's executive ranks.

Loewy was brought in to help style the Letter Series, most likely sometime in 1937. His impact was just as profound as Dreyfuss on Deere—the Letter Series tractors were transformed. The awkward and eye-wateringly ugly sheet metal coverings on the prototypes were gone, replaced with a curved tank and side panels with elegant and simple lines. The combination created an iconic machine that is quite possibly what most people picture in their mind when you say "old tractor."

1939 FARMALL H

This is the fourth Farmall H built, serial number 504. *Kinzenbaw Collection / Lee Klancher*

THE DEBUT OF THE FARMALL M AND H

The new large Letter Series tractors were introduced to the world on August 9, 1939, at Farmall Works in Rock Island, Illinois. The company and the plant worked in conjunction with the local chamber of commerce to put on "Farmall Day," an event that included a parade featuring the tractors and their matched implements, music, entertainment, a luncheon, and a plant tour. More than fifteen thousand people attended.

For the farmer of 100 acres or more, the Model H and Model M were ideal tools. The tractors were powerful jacks-of-all-trades, capable of pulling feed wagons and turning conveyor belts as well as performing the heavy field work of plowing, harvesting, planting, and cultivating.

The work to develop the H and M was some of the finest in IH history. The machines were state-of-the-art, and the matching implements were carefully selected and meticulously engineered. Sales efforts matched the engineering prowess, and sales—particularly post-war, when materials were more easily available—were the highest of any single model in history (with the possible exception of the Ford

BIG NEWS

Farmall Models M and H made a big splash when they hit the farm world in 1939. *University Archives Photograph Collection, Special Collections Department, Mississippi State University Libraries*

N-Series, whose final production figures depend on which variants of each you include).

No matter how one does the math, they sold in staggering numbers. The Model H alone sold nearly half a million units. The numbers highlighted the fact that farmers were turning to tractors like never before, and the Letter Series hit all the high notes. The machines offered cutting-edge engineering and features, a bevy of well thought-out attachments and implements, and the industry's largest dealer network—all driven into overdrive with aggressive, well-funded sales and marketing.

AN INTERNATIONAL AFFAIR

By the time the Letter Series was introduced, IH had a well-established international sales and distribution network as well as manufacturing plants in France, Germany, Sweden, Australia, and Canada. This one is being viewed in Holland on April 25, 1946. *Haren Noman, Theo Van / Nationaal Archief 901-6937*

FARMALL M SN 505

This is the fifth Farmall M off the line, serial number 505. *Kinzenbaw Collection / Lee Klancher*

FARMALL MODEL H

The Model H offered farmers a stylish package that featured all of IH's latest bells and whistles. The Model H and the Model M were based on the same chassis design. Although the two had the same mounting points for implements, they shared very few parts. The Model H was designed to be a two-bottom plow tractor while the Model M was rated to pull a three-bottom plow.

The Model H used modular construction, like the smaller Model A and B. Components could be taken off independently as an assembled unit. This had some advantages when the tractor was being built in the factory and made servicing the tractor much simpler.

The Model H's frame was a hybrid of an integral and beam frame. The rear of the frame was the transmission and final drive housing, while the front was a beam frame that ran from the clutch housing forward to underneath the engine.

The frame design showed touches of Loewy. The engine was rubber-mounted to the front frame channels, presumably to make the tractors smoother to operate. A little less vibration would make a big difference over a long day in the fields.

FARMALL TOUR

The Farmall H was state-of-the-art when it appeared in 1939. *Oregon State University Special Collections & Archives / P040:113*

The Model H's four-cylinder I-head engine used a 3⅜-inch bore and a 4¼-inch stroke and was rated at 1,650 rpm. The engine for the original version was started on gasoline and ran on distillate or kerosene (a gasoline-burning engine appeared in 1940). A variable-speed governor was used. The engine oil was force-fed by an oil pump, and an oil filter strained grit from the lubricant. Air and fuel cleaners were standard as well. An IH H-4 magneto and IH D-10 carburetor were used. The distillate-burning version of the Model H was tested at Nebraska in 1939, where it produced 22.14 horsepower at the belt pulley.

The engine was cooled with an enclosed system that used a radiator and water pump. Cooling was adjusted with shutters that could be controlled by a hand crank. The system was thermostatically controlled, with shutters as an available option for cold-weather use.

The transmission used five speeds, with top gear yielding nearly 17 mph. The top gear was used only with pneumatic tires and was blocked out on steel-wheeled machines. A few models were sold with steel when the tractor was introduced, and a significant number appeared on steel during World War II, when rubber was reserved for war use. A lower first gear and taller fourth gear were made available as optional equipment.

The clutch was a foot-operated Rockford 10-inch single disk. The brakes were external-contracting bands on forged steel drums mounted on differential shafts in housings. They were foot-operated and could be engaged separately or in tandem.

The cotton duck-covered seat found on the smaller Letter Series tractors was standard on the H as well. The drawbar was an adjustable quick-attach model. The turning radius was 8 feet, 4 inches, and ground clearance was about 18 inches. Available options for the Model H included starting and lighting equipment, wheel weights, and rear wheel fenders. A temperature gauge and radiator shutter were standard with the distillate engine, and optional on any Model H with the high-compression gasoline engine. A rarer option was the wide-tread rear axle, which allowed adjustment from 44 to 100 inches.

FARMALL H AND 52R COMBINE

The McCormick-Deering Model 52R combine was produced from 1943 to 1950, with more than 36,600 examples built. The PTO-driven combine was an ideal partner for a Model H.
Tordai Collection / Lee Klancher

POWER LIFTS

Three versions of Lift-All, International's hydraulic lift, were available for the Model H. The regular lift raised the entire attachment in one movement. A delayed lift was also available, which was used when both front and rear cultivators were mounted. The front implement lifted first and the back unit a bit after, so the two effectively came out of the ground at the same point. The third Lift-All was the selective lift, which allowed the operator to raise left and right implements separately.

HIGH-COMPRESSION GASOLINE ENGINE

In March 1940, IH authorized the addition of a high-compression gas-burning engine for the Model H. Even though the gasoline engine was not officially released until 1940, a gasoline-burning Model H was tested at Nebraska in 1939. It produced 19.14 horsepower at the drawbar and 24.28

OLYMPIC FLAME

On January 8, 2010, torch bearer John Nelson carried the Olympic Flame on a 1940 Farmall owned by Barrie Baker. Over the course of 106 days, the torch traveled from Victoria, BC, across the country to Newfoundland, and back to Vancouver. The Farmall stepped in on day seventy-one. Other torchbearers included Steve Nash, Shania Twain, Bob Costas, Jacques Villeneuve, Gordie Howe, and Arnold Schwarzenegger.
VANOC / AM1550-S08-2

DUTCH FARMALL
This machine was photographed in the Netherlands (Holland), where IH had three branch offices in 1939. *Eyssen / Heidemij, Nationaal Archief 160-0947*

1947 FARMALL H
The last Farmall H rolled off the line on January 26, 1953. More than 390,000 were built and sold. *Kevin Blackburn Collection / Lee Klancher*

horsepower at the belt pulley. The tractors sent to Nebraska were likely pre-production machines, and the publication of the gasoline tests could have led some to demand gasoline-burning versions for the showroom floor. A somewhat cryptic note on the company decision (a dated company document ordering engineering actions) to build the high-compression engine for the Model M lends some credence to this theory, although there is no way to be sure.

In August 1940, the kerosene and distillate engine was modified to burn distillate fuel only. A kerosene attachment would be offered for those who wished to burn kerosene. In December 1940, the controls were moved to the rear of the tractor for easier operation. A bracket was attached to the steering post, and the radiator shutter, magneto ignition, and the belt pulley control lever were attached to it. The starting switch was moved to the platform, near the operator's foot, and the starting switch control rod was eliminated.

FARMALL MODEL HV

In January 1942, a new model designed for cane and high-bed vegetable cultivation was released. The Farmall HV used step-down housings and an arched front axle to provide higher clearance. The front end had a whopping 30¼ inches of ground clearance.

The step-down housings provided the arch in the rear of the tractors. Inside the housings, a roller chain transmitted power from the drive sprocket to the rear axle. The housing was sealed, and the chain ran in an oil bath.

The dropped-rear drive used stub axles so that tread width was still adjustable. Additional adjustment was available by flipping the rear wheels, which had offset hubs. If the farmer required more adjustment, optional wheels with varying degrees of offset were available.

The engine, basic chassis, and running gear were the same as the Model H, and the tractor could be equipped with an assortment of special equipment.

1944 FARMALL HV
The high-clearance HV featured larger front wheels, an arched front axle, and a dropped final drive to drive over tall crops. This one was displayed at the Minnesota State Fair's Machinery Hill for many years. *Rolland Wiek Collection / Lee Klancher*

FARMALL MODEL M

The Model M chassis featured a more powerful engine fit in a chassis that used the same design as the Model H. The Model M was wider and heavier, sharing the same mounting points for implements. The Model H and M could interchange implements, but very few parts were shared on the actual tractors.

As on the Model H, the Model M's rear frame, clutch housing, front frame channels, and upper bolster are bolted together to make a rigid frame unit. The brakes (contracting bands on steel drums) and narrow front end were specified in the original decision on the Farmall Model M.

The Model M's four-cylinder I-head engine had a bore of 3⅞ inches and a stroke of 5¼ inches. A variable governor was used, and the engine was rated for 36 flywheel horsepower at 1,450 rpm. The engine was started on gasoline then switched over to kerosene or distillate. The carburetor was an IH model with a 1¼-inch updraft. Oil pressure was maintained with a gear circulating pump, and the oil filter used the Motor Improvement Company's P-20 star-shaped removable filter.

The engine was cooled with a closed system that circulated water through a flat-tube radiator with a water pump. The water temperature was originally supposed to be regulated with shutters in front of the radiator. In September 1939, the company decided to use a belt-driven water pump and thermostat to cool the engine. This change was supposed to be effective on serial number 501, the first Model M off the line,

FARM DEMONSTRATION
These gentleman were the directors at a Farm Security Administration farm near Waterloo, Nebraska. In this September 1941 photo, they gathered for a demonstration of a new Farmall on their acreage. *Marion Wolcott Post, Library of Congress LC-USF34- 059481-D*

1939 FARMALL M

This clean 1939 model spent its working days in Pottawattamie County in western Iowa. It was shipped from the factory to the IHC dealer in Avoca and was sold to a local farmer who worked it for decades. The machine was retired and restored in the county. It was a favorite of the late owner of Farmall-Land, Jerry Mez. *Jerry Mez Collection / Lee Klancher*

PROTOTYPE IH DEALERSHIP

This Farmall M was photographed in a preserved International Harvester dealership in Millville, Pennsylvania. The location was a prototype for the new look the company was experimenting with after World War II, and its blueprint was penned by the Raymond Loewy industrial design firm. See chapter seven for more on how these dealerships were designed. *Lee Klancher*

meaning all Model Ms built should have had a thermostat. "Should" is the key word here, as decisions made in the executive boardroom didn't always make it to the production floor in the desired time. The gas tank held 22 gallons, and an IH H-4 magneto with automatic impulse coupling and integral grounding switch was used.

The transmission was a five-speed unit that used roller and ball bearings throughout except for the reverse idler bushings. The tractor would run at over 16 mph in top gear, which was intended for use only on pneumatic tires. The clutch was an 11-inch single-disk built by the Rockford Drilling Machine Company.

The rear tread width was wider on the Model M than on the Model H, with the M's tread width adjustable from 52 to 88 inches. A wide-tread model increased rear tread width from 52 to 100 inches, although that model is quite rare. Ground clearance was the same on the two models at about 30 inches from the rear axle, and both shared the cotton duck-covered seat.

As tested at Nebraska in 1939, the distillate-burning Model M weighed 6,770 pounds (presumably including ballast) and delivered 24.89 horsepower at the drawbar and 34.16 at the belt pulley.

SCHOLARSHIP WINNER

This 1947 image shows Ansel J. Stubblefield of McLean County Illinois, the winner of a scholarship to the University of Illinois. *Daniel Miles / McLean County Museum of History*

1949 FARMALL M WITH TOKHEIM CAB

This cab is an aftermarket item built by Tokheim Oil Tank & Pump Company of Fort Wayne, Indiana. Some may have been dealer installed, and a patent for the cab was applied for by Robert Hill in 1946. This M lived most of its life in a shed and is equipped with a Joy Rider Level Ride seat and an OEM muffler. *Kinzenbaw Collection* / *Lee Klancher*

FARMALL M WITH ELWOOD MFD

More traction equals better farming, and Elwood began offering kits to add mechanical front-wheel-assist to the Farmall M. The dropbox connects to the belt pulley to drive the front wheels. *Tomac Collection / Lorencz Photography*

FARMALL M WITH WOODEN CAB

The need for cabs was filled by both aftermarket companies and crafty farmers—this example shown in a photo dated December 27, 1941, appears to be a creation of the latter. *Frank Bill / McLean County Museum of History*

GASOLINE-BURNING ENGINE FOR THE MODEL M

In February 1940, IH decided to supply the Model M with a higher-compression engine suitable for burning gasoline. The engine was otherwise the same, with the addition of a new cylinder head that brought the compression ratio up enough to burn 70- to 72-octane gasoline. The intake manifold was designed for gasoline operation, presumably without the fuel-warming system on the manifold for kerosene and distillate. This option was supposed to be available beginning with Model M serial number FBK-18144, meaning it was offered immediately after the decision was made early in 1940.

Interestingly, when the gasoline-burning version of the Model M was tested at Nebraska, it bore serial number FBK-ME533. Either the tractor tested at Nebraska used a prototype gasoline-burning setup, or the note on the decision to offer a gasoline tractor was incorrect. Either way, the gasoline-burning Model M produced a few more horsepower than the distillate machine. The gas-powered Model M produced 26.23 horsepower at the drawbar and 36.07 horsepower at the belt pulley. Later in 1940, the kerosene or distillate version of the Model M was changed to burn only distillate.

FARMALL MODEL MD

In December 1940, the Model MD was authorized. As the name suggests, the Model MD was a Model M with a diesel engine. The engine used the same bore and stroke as the gasoline and distillate Model M engines.

The diesel engine was similar to the D-264 first used in the WD-40 and the TD-40. The engine was started on gasoline in a low-compression mode. Once the tractor warmed up for a few minutes, it was switched over to diesel fuel and high compression via a hand lever mounted on the rear fuel tank support.

The chassis was nearly identical to the regular Model M, with the possible exception of the brakes, which used gray iron drums rather than cast steel drums.

Attachments for the Model MD included an 11-inch belt pulley, PTO, swinging drawbar, muffler, pneumatic tire pumps, wheel weights, spark arrestor, fenders, hydraulic power lift, dual rear pneumatic tires, and a low-speed option.

A handwritten note on the decision specifies that the first Model MD be serial number FDBK-26145. January 13, 1941, is also written on the decision. This was quite possibly the date the first Farmall MD left the factory.

THE MD DIESEL ENGINE

By Jim Allen

Two smaller gas-start, four-cylinder diesels were released for production in 1939, the 248-cubic-inch UD-6 (3.88x5¼-inch bore and stroke). The four-cylinders were available in tractors or as power units. The UD-6 first appeared in the 1940 TD-6 crawler and the WD-6 wheeled standard tractor. The Farmall MD, International's first row-crop diesel, got the UD-6 diesel option for 1941.

These downsized diesels shared many parts with the gasoline and distillate engines of the period and could be built on the same tooling. The new fours were dry sleeved. The UD-6 delivered a maximum of 45 horsepower on the flywheel at 1500 rpm, though the continuous rating was only 39 horsepower. The UD-9 could deliver 63 horsepower maximum and 53 continuous. The UD-16 delivered 100 horsepower maximum and 78 horsepower continuous. These ratings varied somewhat by timeframe and application.

FARMALL MD

The Farmall diesel MD was powered by the UD-6 engine, which was also used in the TD-6, WD-6, and ID-6. It was offered in the M beginning in 1941. The engine was started on gasoline, then switched over to run on diesel. *Wisconsin Historical Society / 161174*

FARMALL MODEL MV AND MDV

In January 1942, the company authorized the Farmall MV and MDV, high-clearance versions of the Model M. The Model MV used the distillate engine from the Model M, while the Model MDV used the same diesel engine found in the Model MD.

A wide front end was used, with an I-beam forged axle mounted in the center to allow it to pivot. The axle was arched to provide additional ground clearance, and the front wheels were larger than those on a Model M. The rear was arched with a dropped-drive mechanism similar to the one used on the original Farmall. The rear tires were of similar size to the standard tires on the Model M. The final drive gearing was changed so that the five-speed transmission drove the MV and MDV at slower speeds. Top speed was about 15 mph on the MV and about 16 mph on the MD.

1948 FARMALL MDV

The high-clearance M diesel was first built in 1942, with the last one driving off the production line in April 1952. *Mecum Auctions*

FARMALL MV

This Farmall MV is at work on a 350-acre farm near Donaldsville, Louisiana. The Model MV is pulling two wagons loaded with as much as seven tons of sugar cane. The tractor shuttled the cane back to the farm, where it was loaded into semi-trailer trucks and hauled to sugar mills several miles away. *Wisconsin Historical Society / 53058*

FARMALL COTTON TRACTORS

By Kenneth Updike

Mechanical cotton harvesters were developed as early as the 1850s, but International Harvester didn't begin to develop its own until Edward A. Johnston proposed a vacuum-powered machine in 1922. That unit eventually failed, but the company purchased a design and tested it in the late 1920s and into the 1930s.

The company rolled out its first production model, the H-10-H, in 1942 and said that more than forty years of development went into the machine. The first successful mechanical cotton picker to be put on the market, and one of those units affectionately dubbed, "Old Red," this one resides in the Smithsonian Institution.

Note that IH also built and sold a Farmall M cotton picker machine, the M-10. This model came equipped with a cotton picker and was first sold in 1943 and 1944, in limited numbers. The model evolved into the M-11—with only a few produced—and the M-12-H, which appeared in 1947.

IH also built and sold very few Model B tractors equipped with a cotton picker, known as the B-10-L. IH also made a C-14 cotton picker that fit the Farmall C. Later pickers could be fitted on the 200, 230, 240, and 404.

International Harvester continued to develop that model through the 1940s, and it evolved from machines that were built on a tractor chassis to machines more like combines with their own unique chassis and power unit.

IDENTIFYING COTTON MACHINES

In addition to machines with cotton pickers installed, IH also offered the Model MCH as a cotton picker version of the Model M. It was set up to have a cotton picker mounted, and the final drive could be reversed (cotton pickers are mounted on the back of the tractor, which becomes the front when picking cotton). The cotton picker MCH models are similar to the high-clear MVs, with some distinct differences. You can tell them apart if you know what to look for.

The most definitive way to know whether you have a high clear or cotton picker model is to look at the tractor's serial number tag. Using a Famall M in this example, a factory-built high clear tractor would have have the prefix "FBKV" on the serial tag. In the IH 350 and 450 models built in the mid-to-later 1950s, the word "hi clear" is printed on the serial tag. Some of the cotton picker models had a "CH" at the end of the serial number. Note that tractors could be retrofitted with a cotton picker, in which case the serial number tag would not have a "CH."[24]

If the tag is missing or is not marked in a definitive way, there are other ways to identify the tractor.

The most obvious part that MV high-clear and MCH cotton picker models share is an enclosed final drive drop box. Note that dropped drives on the M and H were driven by a double roller chain, and on Super Series and later, the dropped

FARMALL H-10-H

The cotton picker was a significant innovation for IH. It spent four decades developing the first successful commercially produced cotton picker, which it announced in December 1942 with the release of the H-10-H. Production was limited due to wartime restrictions. *Wisconsin Historical Society / 78138* drive was gear driven.[25] The axle housing also changed. This is a gear box fitted to each rear axle that raises the tractors rear end ground to tractor clearance. This is needed when growing tall crops like cotton, vegetables, or sugar cane, which can be damaged when struck by a "low clearance" tractor. IH fitted a pair of these drop boxes on these tractors to gain the necessary height to clear the fragile plants.

The front ends are where the differences really lie between cotton and high-clear machines. Cotton picker tractors are fitted with a large single front wheel. They do not have dual front tires or a wide front end. Cotton picker tractors were made this way because they were reverse driven and used the drawbar end of the tractor to act as the "front" or driving side. The cotton picker units were mounted on this end of the tractor too. IH flipped the ring gear/pinion in the final drive on low-drum pickers to have five speeds "forward" when traveling. The seat was reversed as was the steering linkage, brake, clutch pedals, and more.

FARMALL M-11-H

The increased power of the M made it better suited to run a picker. A limited number of M-10-Hs were produced in 1943–44, and another small run of M-11-Hs were built in 1946. The M-12-H was built at IH's Memphis Works facility. Introduced as a 1948 model, it sold well. *Wisconsin Historical Society / 78139*

High clear tractors are fitted with a fixed-tread, high-arched, wide front axle. This axle has auxiliary bracing extending from the front spindles back to a point near the tractor's clutch housing. This bracing was needed to support the axle and prevent it from breaking away from the tractor and causing a catastrophic crash or wreck. The axle itself was cast or forged for extra strength in tough field conditions. This high-arch axle raised the tractor's front end, allowing it to pass over the plants below without damage.

Extra-long spindles gave the axle the added height required to raise the tractor. The front wheel hub/bearings were the same stock items the narrow front Farmall.

Since the high clear and cotton tractors are variations using the same base unit (engine, frame, hood, etc.),it is possible to build a fake or replica unit using spare parts. One method to spot a fake is to look at the letter casting codes IH put on its cast parts. The letters should be the same or very close to each other in alphabetical order on both the tractor and drop boxes.

FARMALL B WITH 1944 ONE-ROW COTTON PICKER

A few of these Model Bs were fitted with one-row cotton pickers. This photograph is dated June 28, 1944. *Wisconsin Historical Society / 49897*

FARMALL H-10-H "OLD RED"

This cotton picker was one of the 1943 early production models, and it worked for decades for the Cotton Oil Company in Fresno, California. The machine was restored and donated to the Smithsonian Institute in 1970. *Wisconsin Historical Society / 11187*

FARMALL MODEL A

While IH was busy reconfiguring its midsized line of machines, the competition was out digging for ways to carve out some market share from the industry leader. One of the markets that had the attention of a number of folks—including IH's sharp-as-a-tack former leader Alexander Legge—was the roughly 4 million farms smaller than 100 acres.

Harry Merritt made his start selling big machines for Hart-Parr in 1911. He spotted the opportunities for a smaller machine and presented drawings to Charles Hart, who famously shredded the drawings and declared that any new Hart-Parr models would be *bigger*, not smaller.[26]

Hart-Parr did eventually try a small machine—the Little Devil—and it was a dismal failure. When Hart was ousted in 1919, Merritt went out the door along with most of the sales staff for Hart-Parr.

Merritt ended up at Allis-Chalmers, and he quickly rose in the ranks, helping the company gain some ground on its red and green competitors. While Allis-Chalmers earned more market share, growing from 2.9 percent of all farm equipment sales in 1929 to 11.3 percent in 1936. John Deere also gained, growing from 20.6 percent to 23.3 percent in the same span. IH lost ground, slipping from 51.8 percent to 44.7 percent.[27] Note that while IH would continue to grow sales, the loss of

FARMALL A

IH's small tractor program had roots in the F-12, when leadership felt the next big thing was a small machine. That program's development was accelerated when IH engineers spotted a prototype Allis-Chalmers Model B in the field. The result was an incredibly useful little machine, the Farmall Model A. This one is being inspected on January 16, 1940, near Bendigo, Australia. *Museums Victoria Collection / 1469046* market share is a trend line that continued. By 1970, IH held an 18 percent market share of North American agricultural equipment sales—8 percent down from Deere.[28]

Allis-Chalmers would not survive, but that was moot when Harry Merritt was driving the team building orange machines. One of his successes was the Model B, a little tractor with a light and innovative torque-tube frame, an arched front axle that offered good crop clearance, a modest 15-horsepower four-cylinder engine, and a price under five hundred bucks. Introduced for the 1938 season, the tractor was a bang-on hit, leaving the industry scrambling for an answer.

Various IH executives had expressed their interest in a lighter version of their F-12,[29] and the emergence of the slick little Allis-Chalmers Model B lit the fires under their efforts. Several IH field men had spotted experimental A-C Model Bs at work, and their missives in May and June of 1937 stated that the model "may give us some competition"[30] and urged a swift response by the IH engineering team.

The red brigade did not waste time. The one-plow 1/F prototype was authorized in September 1937, and by spring 1938, several well-finished Model A prototypes were in testing. The problem they faced was that utilizing a torque tube transmission housing was the logical way to construct a light cultivating tractor. The issue? Allis-Chalmers already had this innovation on the market.[31]

IH contacted Allis-Chalmers, hoping to tiptoe around paying patent licensing fees for the torque tube on the Model A and B, which they knew was all-too-similar to the Allis-Chalmers Model B design. Despite IH's outreach, Allis-Chalmers contested the use of the design, and suggested an appropriate compensation would be for IH to pay A-C several dollars for every Farmall Model A or Model B sold. The battle over this lasted until after the Farmall Model A and B were released to the public. The IH attorneys figured out Allis-Chalmers may have mistimed its patent application and were able to negotiate a modest $35,000 one-time payment to A-C to settle the suit.

An innovation that appeared on these machines was the offsetting of the seat, which gave the operator a better view of the ground. This idea had been around for some time and was tested on the reversible experimental Farmall tractors

SMALL IS BIG

In addition to the push to create a small tractor ideally suited to working with a single-bottom plow, IH invested heavily in the creation of a smaller combine. *Case IH*

SPECIAL DELIVERY

IH pulled out all the stops when introducing the new Farmall Model A, with an elaborate gala at the plant in July 1939 as well as ongoing introductions at dealerships around the world. This is the Horsham dealership in central Australia, which still existed as of 2023. *Museums Victoria Collection / 1469044* of 1921–1922. The little Farmalls combined the offset seat with an entire drivetrain that was shifted, offering an almost unobstructed view of the crops, furrows, and everything else beneath the machine.

Ever the marketing machine, IH dubbed the clever concept as "Cultivision" and splashed it all over the advertisements for its new brainchild. That feature was accompanied by a raft of implements and attachments, all designed or adapted specifically to fit the smallest Farmall to date.

The Model A was the first of the new Letter Series line to launch, and the introduction lifted off with a blast of fanfare. *Harvester World* magazine went above and beyond its usual level of hyperbole to boast of "Harvester's thorough preparation for synchronous invasion of the new market." The initial event was a gala affair held at Tractor Works on July 12, 1939, with an A-list of attendees that included printing and banking executives as well as "technology" journalists, all of them personally courted by IH's top brass.[32]

FARMALL A AND McCORMICK 42 COMBINE
The small farmer focus led to this Model 42 combine, a four-foot unit designed for the millions of farmers working well under 100 acres in the 1940s. This one is shown working near Buffalo, New York, in 1939. *Wisconsin Historical Society / 23689*

The kickoff at Tractor Works was closely followed by dealer previews at regional IH branch headquarters all around the US and Canada. The Harrisburg, Pennsylvania, event organizers handed out Farmall plastic safari helmets. In Toledo, Ohio, they demonstrated the little Farmall towing a rubber-tired wagon loaded with 7,175 pounds of cargo through sandy soil in third gear.

The tractors also represented a union between IH's desire to use Henry Ford's manufacturing techniques and the revolutionary Farmall design pioneered by Bert R. Benjamin and company. The roots of their design harked back to Benjamin's suggestion to build a light tractor like the Fordson that was capable of cultivating as well as plowing. The Model A and B were just that, with enough power and weight for light plowing, and adjustable tread width and great vision for cultivating.

PAIRING OLD AND NEW

In this photo dated June 1940, farmers are harvesting oats with a combine and tractor purchased with a community cooperative loan offered by the Farm Security Administration. *Marion Post Wolcott, Library of Congress LC-USF34-053880-D*

1944 FARMALL A

Painting Farmalls pink became popular with restorers, but it was something that was not done by the IH factory as far as anyone knows. *Megan Loschen Collection / Lee Klancher*

FARMALL AV

The high-clearance Farmall AVs were intended for use with crops with high beds. This image is dated March 21, 1940. *Wisconsin Historical Society / 76469*

More significantly, the two models used the integral frame and unit construction developed for the McCormick-Deering 15-30 and 10-20. The engine, steering gear housing, clutch housing, transmission case, differential housing, and final drive housing were modular pieces. The pieces were then bolted together to effectively become the tractor's frame. The design was strong, efficient, and reasonably easy to service.

Physically, the tractors were small and light. The operating weight of the Model A was just under 1,700 pounds, more than 1,000 pounds lighter than the F-14. The Model A was shorter and a bit narrower than the F-14, as well.

One of the oddest engineering decisions on the Model A and B was the dropped-rear drive mechanism. The simple straight axle and larger rear wheels introduced to IH on the F-12 provided the advantages of easily adjustable rear wheel tread and reduced complexity. The Model A and B used the same dropped-rear drive found on the Farmall and encountered the same problems. Rear tread width could only be adjusted by flipping the offset rear wheels, and the additional parts added weight and complexity. It's possible that IH wanted to keep rubber costs down, as using the mechanisms allowed smaller rear tires to be mounted. Perhaps the added weight of the mechanisms was intended to improve traction.

FARMALL MODEL AV

The Model AV was the high-clearance version of the Model A. It was intended for use in tending to crops planted on high beds that required additional tractor ground clearance. Sugar cane was the most common crop requiring a high-clearance machine, and cane fields used the higher beds to improve drainage.

The Model AV consisted of a group of additions to the Model A. It was authorized for production in October 1940. The package consisted of a new front axle, front wheels, rear axle shafts, rear wheels, drawbar, and steering gear housing base. The result was a 5-inch increase in ground clearance, which brought it up to 27½ inches. The front wheels were 4x19, and the rears were 8x36. The transmission gearing was not changed to accommodate the larger wheels, so the tractor traveled slightly faster in each gear than a Model A. Top speed bumped up about 3 mph to 12 ¾ mph.

Available attachments included a swinging drawbar, wheel weights, belt pulley, PTO, muffler, equipment for burning distillate and kerosene, radiator shutters, electric starting, electric lights, flywheel ring gear, double-groove fan and generator pulley hub, spark arrestor, tire pumps, high-compression gasoline starting, and different compression ratio pistons for higher altitudes.

INTERNATIONAL A

In 1940, the International Model A was introduced. The International was the industrial version of the Model A, and it closely resembled the Farmall Model A. The industrial models were intended for commercial use, and researcher Jim Becker believes that most were used as highway mowers. The tractors were certainly marketed to different audiences, and the International Model A shows up in all kinds of situations—from orchards to factory applications—in archival photographs.

According to Becker, the major differences between the International A and Farmall A were the International A's

INTERNATIONAL MODEL A

This industrial Model A was used by the city of Bloomington, Indiana. The loader and grader blade was used to gather leaves in the fall and to spread cinders on ice-covered streets in the winter. The plow was for snow removal. This photo is dated 1949. *Wisconsin Historical Society / 60009*

foot accelerator and the Model A's optional heavy-duty front axle. The diamond-pattern pneumatic rear tires were more common on Internationals, although they became available for any of the Model As.

The convention of stickering industrial or utility versions as Internationals and the cultivating version as Farmalls eventually became common IH practice. It was unique to the Model A during Letter Series production, but the Hundred Series would use the same convention, replacing the W, I, and O tractors with a standard tread model badged as an International. Orchard and industrial versions could still be ordered, but they no longer had a unique model designation.

FARMALL MODEL B

The Model B was almost identical to the Model A but was designed to cultivate two rows at a time. The single front wheel went between the rows, while the dual front wheels of the Model A straddled a single row. The B also had a wider rear tread width and a tricycle front end. The Model A and B used the same series of serial numbers with different codes.

The rear tread width on the Model B was adjustable from 64 to 92 inches, compared to the 40 to 68 inches of the Model A. The wider rear end made the B slightly heavier (just under 1,800 pounds). The same attachments were offered for the Model B as for the Model A.

POWER LIFTS FOR MODEL A AND B

Power lifts were becoming increasingly popular on tractors by the 1940s, and the Model A and B had several options. One was a simple hand-operated mechanical lift. If the A or B owner wanted a power lift, the optional pneumatic lift was the answer. In many ways, the lift was an elegant engineering solution. It was powered by exhaust pressure, providing plenty of lift power without the drain on engine power of a hydraulic lift. The one major drawback was that if the mechanism got dirty (which tends to happen in farm conditions or after twenty or more years of sitting idle), the lift became less than powerful. Regular maintenance kept the pneumatic

MODEL B

This farmer from Jasper County, Iowa, is adjusting the wires used to plant and properly space rows of corn. The image is dated May 1940. *John Vachon, Library of Congress LC-USF34-060876-D*

1945 MODEL B

The Farmall B was the two-row counterpart to the one-row Model A. The major difference between the two was the narrow front end of the Model B, which was designed to ride in the middle of two rows. The first Farmall B was built on September 5, 1939. *Kinzenbaw Collection / Lee Klancher*

lift working well, but neglect would bring only misery. The Super A would be equipped with a hydraulic system to power the lift, and the pneumatic lift was phased out. Although the hydraulics required less maintenance, the hydraulic system soaked up horsepower. Some say that the Model A has more drawbar power than the Super A due to this.

Other attachments available for the Model A included a swinging drawbar, belt pulley, power take-off, wheel weights, exhaust muffler, equipment for burning distillate and kerosene, electric lighting, a radiator shutter, and several different pistons to change the compression ratio for use at higher elevations. Shortly after being introduced, the Model A received a redesigned rear wheel and a PTO shield.

PNEUMATIC LIFT-ALL AND LOADER

The Farmall A and B could be equipped with a pneumatic lift that was powered by exhaust gas. The lift worked reasonably well when the engine was fresh, less so as the machine wore out. They are now a rare and desired option among collectors. *Wisconsin Historical Society / 73607*

NEW ENGINE PACKAGE FOR MODEL A AND B

In 1940, a high-compression package was authorized for the Model A and B. The company decision reads that the option was offered to meet sales demand and to "provide equipment as tested at Nebraska." The wording insinuates that the University of Nebraska may have tested a more powerful Model A or B than the version IH released to its dealers. Detroit auto makers have certainly been willing to release factory-massaged machines for testing, so it doesn't seem a huge stretch that IH would send a more powerful machine to the Nebraska test. The fact that the decision was not to be circulated to distributors or even the head office makes it likely that IH pulled a fast one on the Nebraska tractor test of 1939.

Whatever the cause, the high-compression package was offered in late 1940 or early 1941. It consisted of a cylinder head, valve guide, valves, seats, intake and exhaust pipe, and all the necessary gaskets, studs, and brackets. No horsepower figures were on the decision, but it's safe to assume that tractors

FARMALL B

Farmer Evan Deirker harvesting castor beans in Havana, Illinois, in 1949.
Bill Frank / McLean County Museum of History

equipped with the package would match the output of the Model A tested in 1939. The output of tractors not equipped with the package is unknown.

FARMALL MODEL BN

In August 1940, a new version of the Model B was authorized, the Model BN. The tractor was simply a Model B with a narrower rear tread width. The tread width was adjustable in 4-inch increments from 56 to 84 inches. According to the factory decision, the parts required to convert a standard Model B to a Model BN were a new left differential shaft, differential housing, platform, and drawbar. The right seat support on the Model BN touches the right fender, while the Model B has a gap.

Collector Jim Becker notes that the Model BN has several advantages over the Model B. For one, fewer were built, making it a rarer model. More importantly, the slightly narrower Model BN fits into his trailer, while a Model B will not.

A swinging drawbar was available as an attachment. A single front wheel was standard, and the more common closely spaced dual front wheels were optional.

The BN was narrow and light for a tricycle design tractor and was a bit tippy. It provided a viable solution for the farmer who worked tightly-spaced crops, but tread width adjustment was still a pain. These problems undoubtedly influenced the development of the Model C, which was heavier, wider, and had the easily adjustable straight axles of the larger Letter Series tractors.

1947 MODEL BN

The Farmall BN was the narrow-tread version of the Model B, with the rear tread adjustable between 48 and 56 inches wide. The need for a narrow-tread model would be more or less eliminated with the Model C, which used a straight axle that was more adjustable than the dropped-rear drive on the Model A and B. *Mecum Auctions*

TRACTOR WARS, ROUND TWO
Ford and IH Square Off Again

A brilliant Irish inventor, Harry Ferguson was the first person in Ireland or the UK to build and fly his own airplane.[33] He accomplished this remarkable feat on December 31, 1909. Not surprisingly, Ferguson was a farm boy. His interests led him to sell American-made farm tractors, and he quickly tuned into the fact that the implements had been developed to pull behind horses and were not efficient when pulled by tractors.

He patented a hydraulic lift system in 1926 and developed a three-point hitch that would transform farming and remains in use today.

Ferguson knew before that time he was on to something good. He met with Henry Ford in 1920. While Ford liked what he saw, the two men could not come to an agreement. Ferguson took his innovation to the David Brown Company and the Ferguson-Brown tractor was produced. Sales were less than optimal, and the relationship between Brown and Ferguson soured. As with so many innovators, Ferguson became dissatisfied with his lack of control over the final product. The men agreed to part ways.

In September 1938, Ferguson once again met with Ford. He demonstrated his Ferguson system in Clara Ford's garden at the family's Fair Lane Estate.[34] The Ferguson tractor outperformed a Fordson and an Allis-Chalmers Model B both in the garden and across the road in heavier soil.

Ford was impressed, and the two men sat down at a wooden table and came to an agreement. The infamous handshake agreement had Ford back in the tractor business—once again "poking sticks in the eyes of the McCormicks" (as Cyrus McCormick III famously wrote in *Century of the Reaper*).

In eleven months of furious development, Ford and Ferguson built themselves a new machine. Ford's chief of styling, Eugene Gregorie, drew the art deco design of the machine (possibly with help from Edsel Ford).[35]

The Ford-Ferguson 9N was introduced on June 29, 1939—a few weeks before IH launched the Model A—and it took a bit of the shine off IH's debut of the Letter Series tractors. The new Ford tractor landed coverage in *Time* magazine, which on July 3, 1939, effused, "The tractor is simple as a motorcar, and can be maintained by . . . any schoolboy."

The Letter Series machines had it all: style, power, famed IH reliability, and the latest bells and whistles. What they didn't have was the Ferguson system.

The Farmalls also faced another issue—the Ford was a bargain. The 9N had just over 23 belt horsepower, meaning it was about the same as the Farmall H. While the Farmall H was priced near $950 in 1939, the Ford 9N was well under $600.

What Harry Ferguson's system did was use a three-point hitch mount with draft control. The system used leverage to press the front wheels down when the plow encountered hard soil. In addition, as the plow hit a rock or other hard object, the rear wheels were pressed down, increasing traction. The system was a tremendous improvement over what was on the market in 1939, and it is still in use on modern tractors.

The concept was much like the one that led to the creation of the Farmall: build the machine around the implement. The three-point hitch became an extension of the tractor. An ingenious system, and a patented one. Ford and Ferguson were the only ones who had it, and they were exploiting it to full measure.

Sales of the Ford 9N were substantial, with more than thirty thousand sold per year from 1940 to 1942. The numbers were about one-third of IH's sales of the Letter Series in those years, but about equivalent to sales of the Model H.

N-SERIES TRACTOR

The Ford N-Series tractor married Henry Ford's sales and marketing might with a brilliantly designed hitch created by Harry Ferguson. The combination was one of the best-selling tractors of all time, the Ford 9N, which appeared just before the launch of the new Farmalls in 1939. Ford was a thorn in IH's side in the first part of the twentieth century. *Ansel Adams, Library of Congress LC-A35-5-M-18-Cx*

HARRY FERGUSON AND HENRY FORD

Harry and Henry examining an early Ferguson model. The two initially had a handshake agreement to build the Ford 9N, a deal that ended in an acrimonious split. *National Museums Northern Ireland HOYFM.L2014.9*

While IH was looking for a hitch system of its own, Ford and Ferguson's famed "Handshake Agreement," in which they decided to join forces to produce Ford tractors with the Ferguson system, fell apart in 1947. Ford jettisoned Ferguson and launched a revised version, the 8N. Ferguson responded with a $340-million lawsuit and his own tractor, the Ferguson. Ferguson eventually won the lawsuit and was awarded $9.25 million.

Ford and Ferguson's antics aside, the Model 8N began selling in big numbers. In 1948, Ford sold nearly one hundred thousand units. In 1949, sales of 8N topped that figure. While IH was still selling significantly more total machines, no single IH model approached sales of one hundred thousand in a year. Ford maintained a rate of sales close to that torrid pace until the mid-1950s, remaining in the domestic market this time. While IH remained the leader, the number of Fords pouring into the market had to bring up bad memories of how close the Fordson had come to driving IH right out of business in the 1920s.

HARRY FERGUSON 1948

In 1946, Henry Ford II announced that the handshake deal between Ferguson and his father was off. Not only was the partnership dissolved, but Ford would build its own tractor. When Ferguson got a look at the new Ford, he promptly sued Ford for patent infringement. They settled out of court in April 1952, with Ford conceding it had infringed Ferguson's patents, would change the design of its machines, and would pay Ferguson $9.25 million. *Mark Kauffman / The LIFE Picture Collection*

STYLED STANDARD TREAD TRACTORS

In 1940, the standard tread line of IH tractors received the styling and features of the Letter Series machine. The machines shared sheet metal, engines, transmissions, and plenty of other parts. Standardization was making its way across the Farmall tractor line.

The W-4 was the standard tread version of the Farmall H and the W-6 was the companion to the Farmall M. This practice cut costs in production and design while simplifying things for the parts man at the IH dealers. Instead of stocking four different sets of piston rings, only two were necessary. This cut the number of common replacement parts they needed to stock by a third or more, adding up to significant space savings on the shelves. More importantly, the styled W tractors were built at Farmall Works, alongside the Letter Series tractors.

Standard treads sold steadily but were not wildly popular. The tractors were well-suited to plowing, industrial, and orchard work. The low height worked well to sneak under orchard branches or through factories. Loewy styling and the latest features of the Letter Series Farmalls made the W-4, W-6, and W-9 solid additions to the IH tractor line when they were introduced in 1940.

McCORMICK W-4

The smaller of the so-called wheel tractors was the Four Series, which included the W-4, I-4, and O-4, all of which were introduced in 1940. If you include the "Super" versions, roughly twenty-four thousand of the Four Series were built. *Case IH*

McCORMICK O-4 AND OS-4 EXPERIMENTAL MODELS

These experimental versions of the orchard four are in original condition. The O-4 has the full coverage fenders—the other is the OS. These experimental models were used to test live hydraulics—which would not appear in production until the Super A in 1947. *Ryan Peters Collection / Lee Klancher*

The gas and kerosene engines for the W-4 and W-6 were taken directly from the Farmall H and M. The WD-6 used the diesel engine from the Farmall MD, and the diesel and gas engines in the W-9 were from the TracTracTor line. The frame differed in that the H and M had front and rear channels that bolted to the block, while the W-4 and W-6 used front and rear gray iron frames bolted together. Mounting pads were cast into the frames, allowing attachments to be hung at several points. The distinction was fairly subtle, but the most important difference was that many of the parts were interchangeable.

Among the more interesting facets of the standard tread tractors were the names. As always, IH was interchanging names to place on the tractors. The W tractors were first McCormick-Deering machines, then just McCormick machines with the new Loewy-designed "IH" logo on the side. Forty years after the merger, the names of the founding companies were still floating around. Actually, this probably had more to do with marketing strategies than anything else. Since the very beginning, IH had been fond of marketing with a wide variety of brand names. While the marketing tightened up over the years, to this day, red tractors are sold with a variety of brand names.

A WORLDWIDE SENSATION

The Four Series machines were powered by IH's C-152 four-cylinder engine, which was rated for 25 horsepower at the belt pulley. This W-4 is shown in Australia in April 1946. *Museums Victoria Collection / 1469674*

McCORMICK O-4

The O-4 was the orchard version of the Four Series. Like the industrial models, orchard models were quite similar to the W-4. The most noticeable feature is the swoopy fenders, designed to keep branches out of harm's way. *Wisconsin Historical Society / 12138*

The orchard tractors were branded as McCormick-Deering early on and McCormick, with the IH logo, later in production.

The industrial tractors—the I-4, I-6, and I-9—bore the International name, and had done so for most of the time. International was used more or less consistently with first industrial machines, and, beginning with the Hundred Series, all of the standard tread tractors were Internationals.

It seems that not as many of the styled industrial tractors were as intensively modified as previous industrial models. The company advertising didn't push custom machines, and photos of heavily altered machines are scarce in the archives.

The standard tread tractors were the first IH tractors to neatly merge wheeled and row-crop production, a practice that became common. The tractors made lower production runs feasible and profitable and were another example of IH ingenuity bringing good equipment to the market.

McCORMICK W-4 & I-4

The W-4 was the standard tread version of the Farmall H and used the same C-152 engine. The four-cylinder unit was started on gasoline and ran on kerosene or distillate. It used a variable-speed governor and was rated for 21 belt pulley horsepower at 1,650 rpm. As with the Farmall H, a high-compression gasoline-only version of the engine was released soon after the model appeared. The transmission was the same five-speed, with fifth being a tall road gear that was locked out when used with steel wheels. The tractor was geared slightly lower than the Farmall H, with a 14-plus miles-per-hour top speed. The usual IH filters for fuel, air, and oil were provided.

INTERNATIONAL I-4

The I-4 was the industrial version of the W-4, and the two were very similar. The industrial version used a foot accelerator. This I-4 is pulling a homemade lumber trailer attachment in Milwaukee, Wisconsin. The sturdy tractor hauled 150,000 board-feet of lumber per day with this setup. *Case IH*

The engine was cooled by a radiator and water pump system, but early models used shutters rather than a thermostat to control coolant temperature.

The front axle was a forged I-beam that pivoted on the front bolster. The steering mechanism differed from that of the Farmall H. On the Four Series tractors, it was a drag link that extended along the left-hand side of the tractor. Forged steel steering knuckles pivoted in bronze bushings in the ends of the front axle. The rear axles were mounted in removable housings. The tractors were reasonably light, at about 3,800 pounds with single rear wheels, and about 5,500 pounds with dual rear wheels. The heavy-duty option, which included beefier front and rear axles and hydraulic brakes, added about 600 pounds more.

The basic standard tread, the W-4, was the first model that appeared. It was approved for production in March 1940. The orchard version, the O-4, was approved shortly thereafter, in April 1940. The O-4 was the same machine with the addition of smooth body work and swoopy fenders that prevented damage to trees as the tractor slipped through the orchard. An underslung muffler and air intake and a hand clutch completed the package, keeping the machine low and compact. Orchard models also have the distinct advantage of stylish Buck Rogers-esque good looks.

The other Four Series was the International I-4, the industrial version of the W-4. The sturdy I-4 could be equipped with dual rear tires (which were, as you probably know, actually quad rear tires), rear work and travel lamps, front and rear wheel weights, and hydraulic remote control.

In 1953, the Four Series went Super, and each tractor was labeled with "Super" on the decal on the front of the hood. Under the hood, the new C-164 engine was up to 164 cubic inches by increasing the bore ⅜ of an inch. Horsepower took a corresponding jump, up to about 30 horsepower at the belt when burning distillate and 33 on gasoline.

McCORMICK W-6

The first of the Six Series machines to appear was the W-6, which was the wheeled counterpart to the Farmall Model M. Just as the Model H and M shared a common chassis design, the Four and Six Series machines were much alike. The significant difference was the engines, especially the diesel engine available in the Six Series. The increased power had the tractor rated to pull a three- to four-bottom plow or an 18-foot single disk.

McCORMICK W-6

The Six Series were the standard tractor versions of the Farmall Model M. This tractor is working with a GL-9A disc harrow near Beveridge, Victoria, in Australia. This photo is dated May 20, 1947. *Museums Victoria Collections / 1469894*

The base model was the W-6, which was authorized for production in November 1939. Production records show that it hit dealership floors in 1940. The tractor used the distillate- or kerosene-burning engine from the Model M, which was the 3⅞x5¼-inch four-cylinder. A variable governor was used, and the engine was rated for 31 horsepower at the belt and 24 horsepower on the drawbar when burning distillate fuel.

The original cooling system used a radiator and water pump with shutters to regulate engine temperature, although

McCORMICK WD-6

The standard tread McCormicks were standardized more than previous models and shared quite a few common parts with their row-crop linemates. *Case IH* later models used a thermostat. As with the Model M, the high-compression gasoline engine was offered as an option in mid-1940 for all Six Series tractors.

Interestingly, in July 1940 IH decided to authorize offering a kerosene-burning attachment for the standard Six Series and Farmall M tractors, despite earlier models being clearly marked as capable of burning distillate or kerosene. It's possible the earlier engines did not burn kerosene well, and the kerosene attachment addressed that problem. All standard Six Series tractors were rated to burn distillate. To burn kerosene, at least according to IH, you had to purchase the kerosene-burning attachment.

The chassis and running gear were similar to those of the Four Series. The transmission was a five-speed, with a tall fifth gear locked out on steel-wheeled tractors. The clutch was an 11-inch Rockford Drilling Company single-plate unit. The brakes were external contracting bands on drums. The Six Series tractors weighed in at about 4,800 pounds with single rear wheels, about 1,000 pounds heavier than the Four Series tractors.

B-25 MITCHELL UNDER TOW

The standard tread McCormicks were put to work in a variety of industrial settings—including during World War II. This battle-scarred unit is towing a bomber near Batavia (now Jakarta, Indonesia) in February 1947. *Nationaal Archief N5-242*

McCORMICK WD-6

The first variant to be introduced was the WD-6, the diesel version. Authorized in February 1940, it should have been on the farm by the harvest season of that year, and in significant numbers by Spring 1941. The D-248 engine was also used in the company's small crawler, the TD-6. The four-cylinder engine used the same bore and stroke as the kerosene/distillate engine and was started on gasoline. Despite the similar displacement, the diesel put out a few more horsepower than the standard W-6 engine, with the WD-6 pumping out 33 horsepower on the belt and 26 horsepower at the drawbar.

The chassis, clutch, and transmission were identical to the W-6. The cost was slightly higher, but the increased efficiency and horsepower of the diesel engine should have made it a popular choice with farmers. No differentiation was made in serial numbers, so production figures are not available.

McCORMICK O-6

The next model to appear was the O-6, the orchard version. As with the O-4, the keys to an orchard model were streamlined fenders and underslung exhaust and intake pipes. The kerosene/distillate engine was used on the O-6, while the OD-6 used the D-248 diesel engine. The O-6 was authorized for production in mid-December 1940, making it available early in 1941. The other Six Series machines were the industrial versions, the I-6 and ID-6, which were introduced in 1940.

McCORMICK NINE SERIES

The Nine Series brought additional power to the farmer and offered the simple improvement of increased capabilities. While the Four and Six Series machines were significant for integrating IH lines, the Nine Series models were important because of their individuality. The Nine Series were big, open-field tractors that used powerful engines from the crawler series and had an integral-frame chassis designed to meet the demand for more powerful tractors. It's no coincidence that these machines were the bestsellers of the wheel tractors; the farmer was demanding more and more power on the farm. A combination of an increase in driven implements, larger implements, and changing tastes made horsepower the key to selling tractors, especially in the 1950s and 1960s.

INDUSTRIAL MODEL I-6

The industrial member of the six family; an ID-6 diesel was also offered. Note that this is likely an early production unit; most of the industrials were badged "International." *Wisconsin Historical Society / 161008*

McCORMICK W-9

The first of the Nine Series, the W-9 was introduced in 1940, with only serial numbers 501–577 produced that year. Like the Nine Series precursor, the W-40, the W-9 used the same engine as the T-9 crawler, the C-335. That engine could be equipped to burn gasoline, kerosene, or distillate, and pumped out 44 horsepower at the belt and 36 horsepower at the drawbar. The diesel engine was actually quite similar, with identical bore and stroke (4.4x5.5 inches) and similar

1947 McCORMICK 0-6

The orchard model six featured a gas engine, relocated controls, and swoopy bodywork. The machine spent most of its working life in an orchard in Stewartstown, Pennsylvania, just an hour's drive south of Lititz. It is shown with the late Ronald Kline and his son, Wesley. *Kline Collection / Lee Klancher*

1947 McCORMICK W-9

The McCormick W-9 was one of the earliest IH high-powered field machines. While its 52 drawbar horsepower isn't much by today's standards, it was one of the most powerful machines in the line when it was introduced in 1940. This W-9 was restored by Bernard Selensky of Baytown, Texas, in honor of farmer Neil Burnside. Selensky worked for Burnside for many years, running W-9s just like this one. Selensky eventually cofounded a seed company. When he sold his portion and retired, his first task was to restore this W-9. He found it in Kansas and did all but the sheet metal work with his own hands. *Bernard Selensky Collection / Lee Klancher*

horsepower output. Diesel Nine Series were indicated with a "D" in the model name (WD-9, ID-9, etc.).

All used the same chassis and running gear. The transmission was a five-speed, with a 12-inch single-plate clutch. As with the rest of the IH line, fifth gear was locked out on tractors with steel wheels. The magneto was an IH H-4, and the carburetor was also an IH unit, a Model E-13. The cooling system used a water pump and radiator and shutters initially, with later models using a thermostat without the shutters.

Optional equipment for the Nine Series machines included a PTO, swinging drawbar, pneumatic tire pumps, wheel weights, and a wide assortment of steel wheels and rims and pneumatic tires.

A Nine Series tractor was the largest, heaviest machine in the IH line, weighing in at about 10,000 pounds with fluids and weights. The weight and horsepower translated into sheer pulling power and traction in any conditions.

McCORMICK WD-9, WR-9, AND WDR-9

Three variants of the Nine Series were built, the regular W-9 and WD-9, the Rice Field Special WR-9 and WDR-9, and the industrial I-9 and ID-9. Perhaps the most unusual of these were the Rice Field Specials. Rice fields are moist and loamy and require tractors with high horsepower and good traction. The Nine Series were well-suited to the task, and the original Rice Field Special was equipped with steel wheels with large spades for traction. Ensuing models used wide pneumatic tires with deep lugs, which must have worked much better than steel.

McCORMICK WDR-9 RICE SPECIAL

The Rice Field Special versions of the W-9 and WD-9 were authorized in August 1944. The original WR-9 and WDR-9 were on steel, probably due to World War II restrictions on using rubber, and used 6x4-inch spade lugs. The Rice Field Specials were equipped with wider rear tires to keep the 7,000-plus-pound tractor from burying itself in boggy soil. *Case IH*

INTERNATIONAL ID-9

The industrial version of the Nine Series was available in gas or diesel versions. All the standard tread industrial machines—the I-4, I-6, and I-9—could be bought in "heavy-duty" versions that were equipped with sturdier rear axles and hydraulic rear brakes. This ID-9 is in the Osceola National Forest near Sanderson, Florida. *Case IH*

STYLED INTERNATIONAL CRAWLERS

When Raymond Loewy was styling International products, the TracTracTors received the same smooth, functional styling as the Letter Series tractors. The first to be restyled was the big fuel-injected TD-18, with prototypes built as early as 1938. The next to be styled were the new T-6, T-9, and T-14. The two smaller crawlers were part of IH's efforts at integrated production, as the T-6 shared componentry with the Model M and W-6, and the T-9 shared parts with the W-9. A few of the machines were introduced in 1939, and the full line was put into production as 1940 models.

All the new crawlers featured replaceable cylinders and main bearings, with the cylinders constructed of heat-treated sleeves machined to fit on the outside and honed and polished inside. The engines were cooled with a thermostatically-controlled radiator and water pump system. Steering clutches and brakes were used for steering, as with previous TracTracTors. Most of the models were available in narrow- and wide-tread versions.

Available options included cabs, electric lights and starting, a variety of track lugs (often called "grousers"), cutaway drive sprockets for heavy mud, radiator shutters for cold weather, and an assortment of guards, bumpers, and shields.

INTERNATIONAL T-6 AND TD-6

The T-6 and TD-6 were the crawler counterparts to the Model M and Six Series. They used the same engines, but the chassis was quite different. The crawler chassis was heavier than the tractors, weighing about 7,000 pounds dry.

The T-6 used the gasoline, kerosene, or distillate engine, while the TD-6 used a diesel unit. Both had a five-speed forward and one-speed reverse transmission, with the top

INTERNATIONAL T-6 CRAWLER

This T-6 was owned by United Airlines and is towing an 81,500-pound DC-6 at Chicago Municipal Airport. It is equipped with rubber tracks to prevent damage to the runway. *Case IH*

forward speed a blistering 5.4 mph. They were available as both wide- and narrow-tread machines, with a 40-inch tread for the narrow model and 50-inch for the wide tread model. When tested in 1940, the gasoline-burning T-6 produced 33 belt and 25 drawbar horsepower on gasoline and 31 belt and 24 drawbar horsepower on distillate. The TD-6 produced 31 belt and 22 drawbar horsepower when burning fuel oil. It was rated to pull a two- to four-bottom plow.

INTERNATIONAL T-9 AND TD-9

The T-9 and TD-9 were the counterparts to the Nine Series. The crawlers weighed about 10,000 pounds and were available in narrow (44-inch) and wide (60-inch) tread widths. Like the smaller crawlers, five-speed forward and one-speed reverse transmissions were used, with a top forward speed of 5.3 mph.

The T-9 gasoline engine produced 41 horsepower at the belt and 32 at the drawbar when burning gasoline. The T-9 was rated to pull a three- to five-bottom plow, and it used the same C-335 engine as the Nine Series wheel tractors. The T-9 was produced only a short time, from 1939 to 1941, probably due to the popularity of the diesel engine.

EXPERIMENTAL TD-18

IH built and tested a number of experimental crawler models. This is the TD-65, a pre-production version of the TD-18, at work with a Euclid wagon on Sky Line Drive in Charlottesville, Virginia, in a photograph dated April 6, 1948. *Wisconsin Historical Society / 7649*

LOEWY'S LINES

This gas-burning T-6 wears its Loewy lines well. The T-6 and TD-6 were smaller machines, weighing about 7,000 pounds and putting out about 30 drawbar horsepower. The original T-6 and TD-6 used the same engines as the W-6 and WD-6. *Raschein Collection / Lee Klancher*

INTERNATIONAL TD-6
The TD-6 was produced from 1940 to 1956, with about 38,500 units sold. This TD-6 is equipped with a Carco winch and a locally built arch. It is delivering pine logs to a small mill deep in the woods north of Martinsville, Virginia. *Case IH*

WAR RELOCATION CENTER
This TD-9 is being put to work by Benji Iguchi in 1943. Iguchi was being held at the Manzanar War Relocation Center, one of several US concentration camps where Japanese Americans were incarcerated during World War II.
Ansel Adams, Library of Congress C-DIG-ppprs-00117

The TD-9 engine was rated for 39 horsepower at the belt and 29 horsepower at the drawbar when introduced. It used the D-335 engine, the same that was used in Nine Series diesels. The TD-9 was produced from 1940 to 1956. During that time, the engine's power output was slightly increased. When tested at Nebraska in 1951, power was up a couple of horsepower at the drawbar and the belt.

INTERNATIONAL TD-14

The original TD-14 was produced from 1939 to 1949. The tractor was significantly larger than the TD-9, weighing about 9 tons with weights and fluids. The TD-14 was available in a wide or narrow tread, with either a 74- or 56-inch tread width.

The original version of the TD-14 was produced from 1939 to 1949. The four-cylinder 461-cubic-inch diesel engine produced 52 drawbar and 62 horsepower at the belt pulley when tested at Nebraska in 1940. The transmission had six forward and two reverse speeds, with a top forward speed of 5.8 miles per hour. The TD-14A was produced from 1949 to 1955.

INTERNATIONAL TD-18

The development of the TD-18 began in the mid-1930s, when the company experimented with a large crawler. At least five experimental models were mentioned in records: the T-80, T-65, TD-65, T-60, and TD-60. The development is discussed in "International Harvester Industrial Power Activities" as follows: "Development of this larger tractor [the TD-18] started early in 1934, when the Engineering Department built the T-80 which had the same rear-mounted steering arrangement as the T-20 and T-40 models. During 1935, we developed and built the T-65 with many improvements but in general along the same lines as the T-80. Then in 1936 we brought out the T- and TD-60 in which the steering clutches were essentially of the

TD-9 AND DISK HARROW

The TD-9 was the next rung up the crawler ladder, weighing in at nearly 10,000 pounds and producing about 40 drawbar horsepower. The original T-9 and TD-9 used the same engines as the W-9 and WD-9. Operated by Lucille Eltiste, this TD-9 is pulling a Dyrr offset disk harrow. *Wisconsin Historical Society / 50918*

INTERNATIONAL TD-14

The TD-14 was good for about 55 drawbar horsepower and weighed over 15,000 pounds. All of the crawlers could be equipped with a wide variety of tread lug, as well as cabs or simple cages like the one on this TD-14. *Case IH*

CRAWLING IN WHEAT COUNTRY

Crawlers were popular with wheat farmers in the first part of the twentieth century. The heavy, powerful machines were good for working the often-expansive wheat fields, and the tracks worked well both in softer soil and on hillsides. Crawlers were particularly popular for farm use in the large, hilly wheat fields of the Pacific Northwest. *Wisconsin Historical Society / 114925*

INTERNATIONAL TD-14A

The TD-14 was upgraded in 1949 to the TD-14A, and it remained in production until 1955. The 461-cubic-inch four-cylinder diesel was good for 63 drawbar horsepower, and the machine weighed in at 18,300 pounds with fuel and accessories. *Lee Klancher*

current production. In 1937 and 1938 this design was further developed and streamlined and was then known as the TD-65."

The TD-18 was larger than any tractor the company had produced previously. It was built in several versions from 1939 to 1957, with each iteration producing more horsepower. The engine was a six-cylinder diesel rated for 85 horsepower on the belt and 70 drawbar horsepower (later versions had 109 belt horsepower). The tractor weighed in at over 10 tons. The tracks were 84⅝ inches long and 18 inches wide, which put 3,046 square inches of track on the ground.

The six-cylinder diesel engine had a bore and stroke of 4¾x6½ inches. The engine featured Bosch fuel injection, full-pressure lubrication, and removable cylinder sleeves and main bearings. The engine was cooled with a thermostatically regulated radiator, water pump, and fan system, with optional shutters for cold weather.

INTERNATIONAL TD-18

The largest crawler in the International line when introduced, the TD-18 was billed as the most powerful crawler on the market. The big machine weighed over 22,000 pounds and the 691-cubic-inch six-cylinder cranked out 71 drawbar horsepower. This one is working with a 42-inch Adams Elevating Grader powered by an IH U-10 power unit in Ledue, Alberta, Canada circa 1939. *Wisconsin Historical Society / 59187*

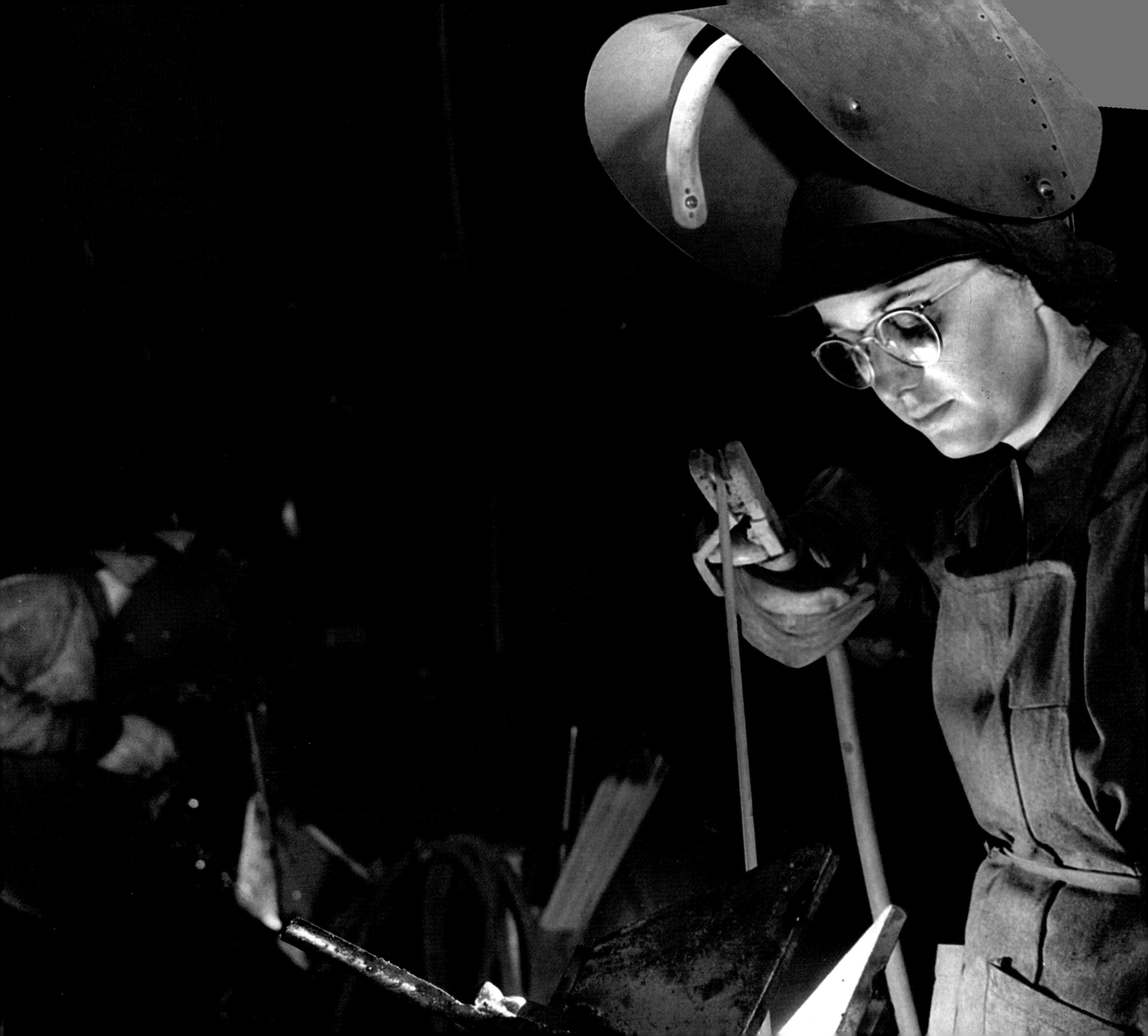

WELDER AT IH FACTORY
Wisconsin Historical Society / 63367

CHAPTER SIX

IH GOES TO WAR

1941–1945

by Jim Allen

"While other women work at machines and in factories—we're soldiers in overalls. . . . We're running the place while Dad's away."
—Toni Taylor, "Women on the Home Front,"
McCall's, May 1942

1941 IH MILITARY TRUCKS

Some of the IH military truck models produced in 1941 are shown here in front of the tower at the Fort Wayne, Indiana, truck plant. From the left, a very early version of the M-2-4 1-ton 4x4, an M-3-4 1.5-ton 4x4, a K-7 2.5-ton dump 4x2, two cab-and-chassis 4x2 trucks supplied to British and Chinese forces, a K-6 cargo 4x2, an M-5-6 cargo 6x6 with winch, a D-30 cargo 4x2, and the mighty KR-8F USMC special 6x4. *George Kirkham Collection*

If there ever was an "all hands on deck" moment for the United States, it was World War II. Individuals and industry responded with an effort not seen before or since. That effort included International Harvester, which built engines, trucks, and construction equipment in addition to its farm machinery. The company's broad manufacturing base expanded during the war as IH was contracted to build armored fighting vehicles, artillery, ammunition, and even torpedoes.

An isolationist trend dominated US politics in the late 1930s, but wise men knew the increasingly savage territorial aggression by Nazi Germany, Fascist Italy, and the Empire of Japan would eventually force the United States into increased preparedness. Our military, especially the ground forces, was behind the times, underbudgeted, under mechanized, and understrength.

A focused US response began when Nazi Germany attacked Poland in 1939, marking the start of its expansionist blitzkrieg across Europe. As part of the US rearmament push, IH started contract work on defense-related projects almost immediately. Things grew deadly serious after the December 7, 1941, surprise attack on Hawaii by Japan. By December 11, Germany and Italy had also declared war on the US. These events triggered a flurry of intense industrial planning that culminated in the government limiting, or outright prohibiting, the manufacture of nonessential civilian items and directing production to war materiel. For the IH truck division, production for civilian use was stopped in March 1942, roughly the same time as most other civilian auto and truck manufacturing was stopped. Farm tractor and implement production continued, but at a slower and strictly regulated pace. Farming was considered essential, but the manufacture of agricultural equipment was regulated and sales rationed.

M-1 ORDNANCE TRACTOR

IH patriotic poster featuring an M-1 ordnance tractor (IH TD-18) towing an M-1 155-mm "Long Tom" field gun from a US Navy LST. *Jim Allen Collection*

WAR MACHINES

As with WWI, farm machinery production was prioritized during WWII. *Wisconsin Historical Society / 27684*

K-1 PICKUP

The US Department of the Navy, which included the Marine Corps and the Coast Guard, became one of International's largest civilian-model truck purchasers. The K-1, K-2, and K-3 models (half, three-quarter, and 1-ton capacities respectively) were built in relatively large quantities. This Coast Guard beach patrol truck has been upfitted with sand tires. *Wisconsin Historical Society / 63422*

CIVILIAN 4X2 TRUCKS IN UNIFORM

All branches of the military needed trucks. Lots and lots of trucks. The earliest purchases were essentially civilian trucks given nothing more than a coat of military branch-appropriate paint. Austere "War Special" trucks became the order of the day. But orders for specific types of civilian trucks continued throughout the war, as they were appropriate to use stateside or at the rearmost military bases overseas.

For IH, the civilian model trucks ordered by the government ran the gamut of gross vehicle weight (GVW) ranges at first. Later, orders focused more on GVWs of 1.5-tons and up. The two IH plants primarily responsible for truck production were in Fort Wayne, Indiana, and Springfield, Ohio. Fort Wayne did mostly heavy and medium trucks and Springfield mostly light and medium, but there were exceptions to both cases.

IH had introduced the K-Line trucks at the end of 1940, so the initial orders of civilian trucks could have been for the older D-Line or the new K-Line. If they were used and standardized by the US Army, they typically had a "G-Number." This was a supply catalog designation from the Ordnance Department that linked to all the parts catalogs and paperwork on that particular vehicle. Most civilian vehicles did not get the familiar "M" designation.

With modifications, certain IH civilian rear-wheel-drive designs were suitable for rough service and hostile climates overseas. This category included some specially designed military trucks with rear drive only. Upgrades typically included brush guards, military blackout lighting, special paint, special tires, special towing apparatus, and tow/tie-down hooks for shipping and recovery.

K-7 DUMP TRUCK

The K-7 line, common in military service, came in four wheelbase lengths. The official US Army designations were truck, dump, 1.5-ton, 4x2, and this IHC unit was designated G541. From 1941 to 1944, IH delivered 6,139 K-7 G541 dump trucks to the US Army. The K-7 chassis was also delivered to the US Army, configured as busses, cargo trucks, tractors, and flatbeds to the tune of about 3,500 units, with more sold to the US Navy, Marine Corps, and civilian government agencies. *Wisconsin Historical Society / 24184*

KR-11 DUMP TRUCK

The 5-ton KR-11 dump was another common military order. The US Army version had a 161-inch wheelbase and was designated G542. It was listed as a dual-purpose truck suitable for hauling dirt, sand, gravel, and general cargo. A total of 3,308 KR-11 dump trucks were delivered from 1942 to 1945. *Wisconsin Historical Society / 63219*

M-425 UNDER STRESS

An M-425 under test on the frame twister at the Fort Wayne, Indiana, proving grounds. The articulation of the truck with a semi-trailer attached is being evaluated here. Many were left behind in Europe after the war and served to jump-start a shattered trucking industry. *Wisconsin Historical Society / 63414*

4X2 CAB OVER ENGINE

The M-425/426 (G671) was a heavily militarized 4x2 cab over engine (COE) truck designed around available parts. With deliveries starting early in 1944, they were just in time to work the Red Ball Express, a masterfully executed supply chain operation that fed supplies to the Allied forces as they raced eastward across Europe after the D-Day invasion. *Wisconsin Historical Society / 26125*

USMC KR8F

Early in the war, the USMC needed a heavy truck. IH had already become their go-to truck source, and in 1941 it supplied a fleet of modified KR8F trucks. The KR8F was a beefed up 6x4 rated for 8 tons, and it featured the Hendrickson tandem drive rear suspension, a dump bed, brush guard, blackout lighting, and an unusual traction/floatation device for the rear that turned the rear tandem duals into pseudo-tracks. *George Kirkham Collection*

FULL TACTICAL FRONTLINE AWD TRUCKS

As the US began rearming in the late 1930s, International Harvester had little direct experience with all-wheel-drive (AWD) vehicles. Through today's lenses, we might see that as an oversight, but viewed in context for the era, it was perfectly normal. AWD was not common. Prior to the rearmament push, AWD vehicles were only being built in small numbers for specialty commercial or military customers. That wasn't because the technology was overly complex, but because small demand resulted in a prohibitively high price and, in light of the Depression, a general market never developed.

From the commercial side, some of that small demand was being fed by companies like Marmon-Herrington, who converted rear-drive trucks to AWD by adding transfer cases and driving front axles. Larger truck manufacturers like Corbitt, Dodge, FWD, GMC, and Indiana built small numbers of specialty AWDs as needed for military or civilian buyers. The even smaller potential market for light-duty 4x4s was not addressed until the mid- to late-1930s when Marmon-Herrington began converting Ford light trucks and cars to AWD.

This all changed in 1939 with the rearmament programs. IH began developing and testing military AWD drive trucks in 1940 in all GVW categories. They coalesced into four, the M-1 Series half-tons, the M-2 Series 1-tons, the M-3 Series 1.5-tons and the M-5 Series 2.5-tons, the capacities being the

M-14 PROTOTYPE
The 1940 prototype IH 4x4 light truck was the predecessor of the M-1-4. The haphazard appearance of the body indicates this was a test mule, but it got them across the procurement finish line. Few specifics on the build are available, but it likely used similar, or the same, running gear as the production M-1-4. *Jim Allen Collection*

M-1-4 CARGO TRUCK
Only 1,123 M-1-4s were built and the majority were used as ambulances. Dave Klink's restored M-1-4 cargo truck is one of the earliest remaining survivors, the 133rd built in fact. It's also noteworthy in being the 63rd with the uprated 233-cubic-inch Green Diamond engine. The M-1-4 was originally fitted with 7.50-16 tires, but Dave's has been updated to 9.00-16s, which was also common back in the day. *Dave Klink*

EARLY PRODUCTION M-1-4 CHASSIS

Most half-ton chassis were allotted for conversion into ambulances. Boyerton built ambulance bodies, and this is how the trucks were shipped from IH for conversion. This is an early production truck, as evidenced by the six-lug, split-ring wheels. It would have used the 214-cubic-inch flathead six as well. *Jim Allen Collection*

off-road load ratings as categorized by US ground forces. On road, the capacities doubled.

Once testing was complete, IH submitted trucks to the service branches for tests. There was plenty of competition from other manufacturers in all categories, but winning a military contract was not a side-by-side torture test of competing vehicles where the "best" was selected. Nothing could be further from reality.

The process usually started with individual formal requests or general invitations to bid from a government agency based

M-2-4 TRUCK

The M-2-4 1-ton became the standard light 4x4 truck of the USMC. It mounted the 233-cubic-inch, 93-horsepower Green Diamond flathead six backed up by a four-speed manual and a two-speed transfer case. Early M-2-4s had six-lug Budd wheels and 7.50-16 tires with rear duals, but most were issued with five-lug "combat wheels" and 9.00-16 tires. *Wisconsin Historical Society / 111625*

ROCKET DETACHMENT

The M-2-4 could be a potent weapons platform. While it occasionally mounted light and heavy machine guns, the addition of the Mark-7 launcher and 4.5-inch barrage rockets made it a formidable weapon. Shown here is part of a rocket detachment in the battle for Iwo Jima in 1945. *Wisconsin Historical Society / 8697*

IH MILITARY PRODUCTION LINEUP

The production heft of IH is illustrated in this image, with lines of M-3L-4 trucks surrounded by crawler cranes and what appear to be M-3H-4 trucks to the right. The US Navy Seabees used the majority of M-3-4 trucks, but the open cab trucks often saw service with the USMC. *Wisconsin Historical Society / 63778*

on preliminary specifications. The manufacturer submitted one or more test prototypes built to those specifications. Those vehicles were evaluated to see how closely they followed the design parameters and tested for strength and reliability. They didn't have to be perfect, they just had to show good potential. If they met the criteria, it came down to the lowest bid, balanced by the government's assessment of how well the manufacturer would be able to fulfill the production terms of the contract in a timely manner. By all reports, IH passed all tests but the bidding. On the huge contracts for the standardized 2.5-ton 6x6s, 1.5-ton 4x4s, and half- to three-quarter-ton 4x4s, GMC, Chevrolet, and Dodge won the respective bids—but IH was not out of the game.

In an ideal scenario, all US armed forces would have used the same standardized vehicle designs. That wasn't possible for many reasons, primarily because the amount of equipment required was so large that even big companies like Dodge, GMC, and Chevrolet couldn't supply enough for all branches of the service in the required timeframes. Nor was it practical for another manufacturer to retool and build

M-5H-6

The upgrade to the closed cab Lend-Lease order was the long wheelbase M-5H-6. They were updated with an open cab, a special cargo body to suit USMC requirements, Thornton Locking Differentials in the tandem rear axles, and upped tire sizes from 7.50-20 to 8.25-20. This truck is considered the best performing 2.5-ton 6x6 of WWII. *Wisconsin Historical Society / 63909*

those companies' designs. A further complication was Lend-Lease, a new program to supply the Allies with hundreds of thousands of necessary vehicles. It is in these alternative areas where IH found a place.

IH was tasked with being the primary supplier to the US Department of the Navy, which included the United States Marine Corps (USMC), thus standardizing and simplifying its logistics mostly to one manufacturer. The USMC would refine its needs, and IH would alter its original designs to build AWD trucks for the USMC in large numbers throughout the war. On top of that, IH answered the Lend-Lease call, sending most of its products to the Soviet Union or Britain.

FIRST PRODUCTION M-5 SERIES

The first production iteration of the M-5 Series 6x6 is shown here on the Fort Wayne Proving Grounds frame twister. The initial build of five hundred trucks took place at the Fort Wayne Truck Plant in 1941. The Hendrickson tandem rear axle suspension was known for being flexy, as illustrated. *Wisconsin Historical Society / 63206*

WRECKER TOWING M-2-4

The basic M-5H-6 was used in many different roles beyond a cargo truck, such as an artillery prime mover, shop truck, fuel truck, water truck, and, as shown here, wrecker. At what looks to be Camp Pendleton, California, this wrecker is towing a rather battered M-2-4. *Wisconsin Historical Society / 63893*

M-5 HALF-TRACK PERSONNEL CARRIER

The M-5 half-track (right) was the first type built by IH, and it mimicked the M-3 half-tracks built by White and Autocar. It could carry a total of thirteen troops and had a single mount for a .30 caliber machine gun. *Wisconsin Historical Society / 23475*

M-9A1

Shown below is an M-9A1 with the armored pulpit for mounting a .50 caliber heavy machine gun. This was primarily designed for anti-aircraft defense but was often used in a fire support role against enemy troops. *Wisconsin Historical Society / 102212 (left) / 102210 (right)*

TRACKS AND ARMOR

International Harvester was not only involved in the logistics of WWII, it also produced fighting vehicles. Early in the rearmament process, the US Army settled on the armored half-track (wheels up front, tracks in the back) as a multi-role infantry support vehicle. It had enough armor to protect the vehicle from small arms fire, and the rear tracks and driving front axle gave it the off-road mobility to keep up with tanks. It also had a sizable carrying capacity and could be used as a weapons platform. Half-tracks carried everything from light machine guns to multiple anti-aircraft gun mounts, mortars, and Howitzers from 75 to 105 mm. Early in the war, mounting a 3-inch anti-tank gun on a half-track proved to be an effective, if interim, tank destroyer.

American experimentation with the half-track began in the early 1930s and culminated in 1939 with an armored design from White Motor Company that became standardized. Like many vehicle manufacturers, White could not supply all the demand, so production was farmed out to a number of other manufacturers, including IH. Each manufacturer followed the established pattern and used the same track system, but the power trains and details of construction were unique to each manufacturer. IH began planning to

build half-tracks in April 1942, and production commenced the following October.

The IH half-track used the biggest engine in the bunch, a 451-cubic-inch Red Diamond OHV six rated at 141 horsepower and backed up by a four-speed transmission and a two-speed transfer case. Besides some slight differences in

ANTI-AIRCRAFT M-17

The M-17 multiple gun carriage mounted four .50 caliber heavy machine guns in an M45 Maxson power traverse and elevation turret. Designed for anti-aircraft defense, it could hurl up to 2,200 rounds per minute into the air. *Wisconsin Historical Society / 63418*

appearance, the IH half-track also used a bit more armor than the other makes, $\frac{5}{16}$ inch versus $\frac{1}{4}$ inch, and it was welded rather than bolted. That made the IH half-track heavier and a little slower.

The IH half-tracks came in six models, the original M5 personnel carrier, the M5A1 personnel carrier that featured an armored pulpit with a ring mount for a .50 caliber heavy machine gun, the M9 personnel carrier that was an update to the M5 with a rear door and updated seating, the M9A1 that added the armored pulpit and heavy machine gun, the M13 multiple gun carriage, and the M17 multiple gun carriage. Most IH half-tracks went out for Lend-Lease to the Allies, the British and the Soviets in particular. IH produced a total of 13,622 half-tracks of all types before production ended in March 1944, all having been built at the Springfield, Ohio, truck plant.

The elephant in the Harvester room was the M-7 tank. Beginning in 1941, development began at the Rock Island Arsenal US Army facility for a new upgunned light tank to replace the prewar M-2 and M-3 Series. The new light tank was given the development moniker T-7. There was the usual series of developments and changes, and a number of prototypes were built before a design was finalized. The result was

M-7 MEDIUM TANK

The near final version of the IH-built M-7 medium tank. Some of its advantages over the standard M4A1 Sherman of the time included a lower profile (smaller target), an arguably better suspension system, and a rail system for the engine that allowed very rapid power plant changes. *Wisconsin Historical Society / 24234* a tank too heavy to be called a "light tank," so it became the medium tank, M7. In January of 1942, IH was handed the preliminary blueprints from these developments and walked away with a contract to build three thousand M7 tanks.

IH decided to build the tanks at a facility in Bettendorf, Iowa, not far from the Rock Island Arsenal. Thirteen preproduction tanks were completed in late 1942, with around half going to various agencies and locales for testing. While there was nothing wrong with them, they were generally inferior to the standard M4 then in mass production, and the wise decision was made to discontinue the project so as not to pollute the supply chain. A total of seven were accepted and delivered in 1943, only to be declared obsolete soon after. One survives in storage at the US Army Center for Military History.

IH TANK PLANT

Shown here is the Bettendorf, Iowa, IH tank plant around the time M7 production was canceled. Though it's often stated that only thirteen (according to the most reliable source) of all types were built, and only seven accepted and delivered, you can see a whole lot more in progress that were no doubt scrapped before completion. *Wisconsin Historical Society / 24240*

CRAWLERS GALORE

By the start of WWII, International Harvester was already a big player in the agricultural and construction crawler markets. It had a wide range of highly regarded tracked tractors in production when the US entered the war, and military and civilian government agencies took full advantage of that. At first, civilian tractors were purchased, but later the types, sizes, and uses were standardized for the branches, and IH updated its products to better suit military needs.

The military broadly categorized crawlers three ways, ordnance (artillery) or general-purpose crawlers, crawler cranes, and bulldozers. IH built diesel- and gas-powered crawlers and, generally speaking, the US Navy/USMC preferred diesel engines and the US Army preferred gas engines for units operated in theater.

Traditional low-speed crawlers were broadly categorized by capability—light, medium, or heavy—and given an M-number. The light, medium, and heavy categories would each have an M-1 version, but they could be from several manufacturers. In the M-1 heavy tractor category, for example, were tractors from International Harvester, Allis-Chalmers, and Caterpillar, each different but with standardized military features such as towing apparatus, seating, and lighting. A military unit could

M-5 HIGH-SPEED TRACTOR

This is the initial version of the M-5 high-speed tractor. Note the fuel tanks are above the tracks and the passenger compartments are divided front and rear. For the majority of production the fuel tanks were moved below the floor. *Wisconsin Historical Society / 24146*

M-5 PULLING 155-MM HOWITZER

The M-5 was used to haul medium-weight artillery over rough terrain, typically the 105-mm Howitzer, 4.5-inch gun, and 155-mm Howitzer (shown here). The tracks gave it high mobility, and it could attain 35 mph on improved roads. It had a long career in the US Army through Korea and into the mid-1950s. *Wisconsin Historical Society / 24690*

TD-14 CRAWLER

The US Army Air Corps specified a certain crawler crane, the M-3, which used a 2-ton Hughes-Keenan crane. It was mounted on an IH TD-14 crawler, which was in the medium crawler class. It was powered by a four-cylinder diesel displacing 461 cubic inches and making a rated 68.5 brake horsepower. *Wisconsin Historical Society / 6592*

USMC TRACTOR CRANE

Both the US Army and the USMC used a version of this tractor crane based on the IH T-9 gas or TD-9 diesel crawler. Both featured a 2-ton crane from Trackson. The USMC version (shown here) had a diesel engine that displaced 334.5 cubic inches and made 44 belt horsepower. *Wisconsin Historical Society / 64115*

be issued an M-1 heavy tractor of any make, though there was usually an attempt to issue one brand within a general operational area to ease parts logistics problems.

Before the war, large artillery pieces were towed by heavy trucks on improved roads and by crawlers in rough terrain. Trucks were needed because crawlers had top speeds under 10 mph, so timely transport often required switching back and forth between truck and crawler according to the terrain. Besides the truck and crawler, another truck and trailer was needed to haul the crawler over the fast portions so it arrived at the rough terrain at the same time as the gun.

The answer to this inefficiency was found early in the war, with a series of unarmored high-speed tracked tractors that used the running gear of tanks. That development resulted in the 13-ton M-5, 18-ton M-4, and 38-ton M-6 high-speed tractors, each type built by a different manufacturer. IH was approached to build one category in August 1941, and the result was the very successful and highly regarded 13-ton M-5 high-speed tractor. It was not originally designed by IH, though they added some refinements. It became a mainline unit for the US Army, putting IH on the military map.

US ARMY M2

When a light crawler was needed, the US Army would sometimes use the M2. Three manufacturers filled this category, the Cletrac AG, the Caterpillar 20, and the IH T-6 (G113). The T-6 uses the same engine as the Farmall M, and like the M-Series, there was a diesel version, the TD-6 crawler. *Jim Allen Collection*

TD-9

The IH TD-9 was considered a light tractor, as was its gas-powered counterpart the T-9. This TD-9 is in use by the USMC, unloading what appears to be a 90-mm anti-aircraft gun from a landing craft at Rendova, in the Solomon Islands, in 1943. *Wisconsin Historical Society / 12148*

US ARMY M-1

In the M-1 heavy tractor class, the mighty IH TD-18 was a heavy hitter. At the start of the war, it was the biggest IH crawler. A US Army M-1 is shown here in 1941, set up as an ordnance tractor, towing the M1918 155 mm gun, which could punch out a 95-pound, high-explosive shell over ten miles. *Wisconsin Historical Society / 6630*

TD-14

The TD-14 was put into the M-1 medium class, but it's putting on a heavy-duty show on this Pacific Island beach in 1945. It's towing two 2,000-pound ammo pallets behind and carrying another in an extension on the dozer blade. *Wisconsin Historical Society / 64031*

FARM TRACTORS AT WAR

It would be hard to argue that farming wasn't vital war production. That's why tractor production continued throughout WWII and farmers were given exemptions from the draft (though many joined anyway).

Prior to the war, the government encouraged a reduction in crop production due to a perceived glut of certain commodities. That changed shortly before the war, and the resounding credo became "grow, grow, grow!" Food was also part of the support packages sent to allied nations, and by the end of the war about 13 percent of US farming commodities were being sent overseas.

Farm production grew increasingly difficult as draft-related labor shortages came to the front. This resulted in women entering the farm labor force (Tractorettes), Mexican nationals being brought over the border (the Bracero Program), and German and Italian prisoners of war being hired out to work on farms. The Geneva Convention did not allow forced labor, but POWs could be hired, and many returning Axis POWs took home a vital nest egg of US dollars that helped restart the European economy.

Farm equipment was strictly rationed during WWII, and many historians think that slowed the transition of small American farms from horses to tractors. Many farmers had to formally request a waiver to replace existing farm equipment, to buy new equipment, and sometimes even to get spare parts.

But tractor production continued, with some of it going to government or military service. It's known that many, if not most, were largely unaltered civilian tractors, sometimes with a coat of military paint and sometimes with just a government number to note its service. Not many received special military modifications.

ID-9

The ID-9 was an industrial tractor in the same category as the standard tread WD-9 diesel and W-9 gas farm tractor. This unit was destined for the US Army and the image is dated 1941. Though the image caption does not indicate this tractor's purpose, with dual rear tires of a street type tread, use as an aircraft tug or something similar is a strong possibility. *Wisconsin Historical Society / 59490*

AI TRACTOR

The year is 1942 and an Industrial AI tractor equipped with turf tires and a mower is seen working at the USMC base at Camp Pendleton, California. The female civil service operator is noteworthy, highlighting the role women played filling in for the men who had been drafted. *Wisconsin Historical Society / 64015*

I4 INDUSTRIAL

The I4 Industrial shown here in 1942 is working as an aircraft tug, towing a B-25 Mitchell bomber built by North American Aviation at its Kansas City factory. Owned by North American, technically it's not a "military" tractor, but it's performing a military service and likely pulled many of the 6,600 B-25s built there between 1941 and 1945. *Library of Congress LC-DIG-fsac-1a35288*

HEBARD A-14 SHOP MULE

The US Navy would sometimes use IH tractors as flight deck tugs on aircraft, with W. F. Hebard & Company as an intermediary company in the production of these tugs. Shown here is a Hebard A-14 Shop Mule on the flight deck of the USS *Lexington* in 1943. *National Archives*

LA VALLE, WISCONSIN, FARMER
Wisconsin Historical Society / 58956

CHAPTER SEVEN

THE POST-WAR BOOM

SALES AT THE HELM 1946–1953

"They hark back to a remembered past when land was the only measure of wealth and one's people owned a quarter-section instead of a body shop and, by God, their name meant something in these parts."
—John Hildebrand, *Mapping the Farm*

Ed O'Neill grew up riding his horse, Blue, to Burr Oak School near his family's farm just south of Rochester, Minnesota. The horse stood 14 hands tall, and Ed was small for his age. He loved to ride the big horse bareback, at times performing tricks he saw in the movies, such as hanging under the saddle, while exploring the area on the fastest form of travel available to him.

His family farmed the rich, hilly land of southern Minnesota, where the farms are overshadowed by one of the world's premier medical facilities, the Mayo Clinic. The clinic draws patients from around the world for advanced treatment of cancer and much more and pulls young talent from the farms by providing good paying jobs at the hospital.

When Ed went to high school, he studied agriculture, learning about animal husbandry and hybrid seeds. In 1937, utility poles made their way to his family's farm and he and his siblings no longer had to milk the family's thirty-three cows by hand under smoky kerosene lights. The new power gave them a Rite-Way electric milking machine and bright new electric lights.[1] His family was on the cutting edge of the rural electrification movement. In 1940, only 33 percent of American farms had electric power. That would shoot up to 78 percent in 1950, and 88 percent in 1952.[2]

When Ed graduated, his brothers ran the farm and he took work as a hired man to the neighbors, shucking corn, driving tractors, and even loading chicks into incubators. It wasn't a career, but it passed the time.

Two weeks after Pearl Harbor, Ed and a friend wandered west to California, finding work building B-24 bombers in San Diego. After watching P-39s land at a nearby field, Ed and his

HARVEST

Bringing in corn in Queensland, Australia, in a photo dated May 10, 1950. *State Library of Queensland / 189512*

friend enlisted in the Air Corps. His daring on the horse transferred readily to the air, and he became a pilot, flying P-38s with an outfit in the 475th fighter group known as Satan's Angels on Biak Island, near New Guinea. His group relocated to Clark Field in the Philippines as the war was winding down.

He returned to the farm for a bit, reenlisted for a boring year flying trainers in Denver, and then came home, got married, and stayed on the farm for good. The GI Bill gave him access to regular ag classes, and his education in high technology farming continued.

After he returned, the farm changed quickly. With fewer horses, the oat fields that fed them were replanted with soybeans that brought in cash. A new four-row planter and more powerful tractor allowed him to plant twice as much corn as his father used to do.

The tractors available at that time gave farmers access to an unprecedented level of technology. Engine reliability, ease of use, and performance had improved dramatically since the 1920s, and the addition of hydraulics, power take-offs, more efficient hitches, and pneumatic tires offered tremendous yield and time-saving improvements.

Combined with the exodus of help from the farm, tractors reached the zenith of demand in the 1940s. In 1920, only 3.6 percent of American farmers owned a tractor. By 1940, 23.1 percent of American farmers had a tractor. By 1950, roughly half of American farmers owned one.[3]

The number of farms in America surpassed 6 million in 1940.[4] That dropped to 5.3 million by 1950.[5] If one-quarter of American farms bought a tractor for the first time in that decade, you had 1.25 million machines going out the door. Add in the number of farmers who upgraded to new tractors, and you have millions of units being sold.

That is, in fact, just what happened. IH sold tractors in unprecedented numbers after the war—and the sheer volume of large tractors sold will never be replicated. The scaling of the company during this era is incredible—IH invested more than $70 million in new plants, equipment, machine tools, and distribution facilities between 1945 and 1948.[6] That is $850 million in 2023 dollars. They also increased staff by 40 percent, and production capacity by 70 percent.

The increase in capacity was accompanied by record-setting sales in 1944, 1947, 1948, and 1951. The growth was

MANUAL LABOR

The advances of the late 1940s would greatly reduce manual labor. Electric milkers, for example, replaced hand milking, which is what Pat McRaith and his niece are doing here to the seventeen cows on his 160-acre farm in Meeker County, Minnesota. *John Vachon, Library of Congress LC-USF34-064451-D*

WINTER MORNING

Pictured here is an early morning milk pickup at the McRaith farm in February 1942. *John Vachon, Library of Congress LC-USF34-064440-D*

YOUNG FARMER

Mike McRaith farms 80 acres in Meeker County, Minnesota, in a photo dated February 1942. *John Vachon, Library of Congress LC-USF34-064441-D*

mind-boggling—IH went from $443 million in total sales in 1943 to $1.28 billion in sales in 1951. Already a behemoth of a company, IH more than doubled its income in this incredible era.

On the farm, the new machinery combined with better seeds and farm techniques had yields up across the board, along with costs. Government programs were confusing and contradictory. Combined with declining profit margins, more and more young Minnesota kids left the farm for good-paying jobs in Minneapolis or at the hospital in Rochester.

In years past, Ed's father, John, had purchased every piece of land he could afford. This gave them 500 acres of ground to work—a sizable farm at the time, but it was split between the three brothers' families. Between 1920 and 1950, average farm sizes grew nationwide from 148 to 215 acres, but the growth differed by region. In the northern plains, average farm size grew from 359 to 491 acres.[7]

Overall acreage being farmed also increased steadily in the late nineteenth to early twentieth century, growing from 250 million acres in 1850 to just shy of 1.2 billion acres in 1950. That growth rate was very rapid from 1940 to 1950, with 200 million new acres added to the total land being farmed.[8]

Ed bought another piece of land after the war—227 acres situated on both sides of Partridge Creek, where it funneled into the Root River. With high bluffs and steep banks, the land had little to offer in the way of farming beyond a few modest fields. The attraction on the property was timber. The family cut and hauled yellow elm, basswood, red oaks, and hard maple off the land, curing some of it to build one of the brothers a new home.

Farmers of the day began a practice that would grow over time: finding secondary means to supplement farm income. In fact, between 1930 and 1950, the number of farmers who had other primary sources of income rose 40 percent.[9] Some, like the O'Neills, logged. Others ran heavy equipment or took day jobs as lineman and factory workers, or even commuted into the city for office work. Farming was becoming less profitable, and people began to mix and match income sources to make ends meet.

Farmers weren't the only ones struggling with profit margins. Despite record-breaking sales, IH continued to record low single-digit profit margins. The company continued to pay high dividends—a practice that kept investors happy, indicated strength to the market, and was often done to maintain or increase overall company value. But the profits after the dividend payout were hopelessly thin. Calculating profit margin by using the figures in the IH annual reports show 1 to 4 percent margins from 1940 to 1952. The two most comparable competitors in the era, Caterpillar and Deere, ran much higher margins, at times close to 10 percent.[10] This was a problem that would not go away for IH.

IH also chose to attack two competitive and saturated markets—industrial equipment, which Caterpillar dominated, particularly in crawlers, and home appliances. The investments in these areas—and a lack of expense dedicated to developing IH's farm equipment—were the first of a series

of key errors made in the 1940s that would haunt the company for the next four decades.

Back in Minnesota, Ed O'Neill ran his farm through the 1940s, with new equipment giving him just enough yield and efficiency to stay afloat. Fires would burn his crops, the local transportation department would try to run a superhighway through his land, and government policies would continue to offer assistance one year and frustration the next.

On a hot day of baling in July 1954, a neighbor touched down on Ed's fields with a Taylor-Craft high-winged plane. He took Ed for a ride and turned over the controls. Ed found his war-time skills with P-38 hadn't left him, and he spent the afternoon happily touring his land and dipping his wings. Easy times were never part of farming, but some days were good.

APPLIANCE DELIVERY

IH looked to serve farmers with a line of appliances. It found dealership sales tough and tried to sell them from these delivery trucks. *Wisconsin Historical Society / 70204*

FARMALL CUB

Opening new markets was crucial for IH, especially in the postwar period. The only way to keep moving machinery was to enter new niches, develop radical new machines, or sell tractors to people who didn't already have them. The latter is the path of least resistance, and IH was well aware that on more than three million farms, farmers still didn't own tractors.

Small farms of less than 50 acres did not typically have a tractor. Some of them were vegetable, tobacco, and livestock farms. Many were worked by part-time farmers who had another source of income. The existing F-12 and Model A targeted small farms of less than 100 acres. The Cub took that a step further, going after even smaller farms.

Some researchers and enthusiasts are fond of saying that the Cub was built for tobacco farmers. Although tobacco farms were typically around 25-acre operations, they were only a small portion of the market. IH expressly stated the Cub was designed for the 2 million American farms of 40 acres or less with a gross income of $400 or less.[11] It was intended to be small and cheap enough to lure new buyers who perhaps dreamed of a tractor for their little operation but found previous models way too big and expensive to be practical.

The Cub was seen as early as October 1945, when it was one of the star attractions at a show at International's experimental farm in Hinsdale, Illinois.[12] That event was the first time in company history that experimental machines were put on display, and IH had sixty-five implements, tractors, and crawlers out on the grounds. IH stated it did this to showcase not what it had available, but what was coming that would "enable the farmer to produce more at less cost."[13]

IH added it would not be able to build the Cub as its existing facilities were at capacity, promising to build a new plant at "Wood River Works." That plant never materialized; instead IH purchased a plant in Louisville, Kentucky, to build Cubs.

IH wrote of the Cub, "probably no other machine in the demonstration possesses greater potentiality of service and usefulness than the experimental Farmall Cub." The company listed truck farms (small farms that sell vegetables at local

THE UNVEILING

In an unprecedented move, IH unveiled several experimental machines not ready for production at a show held at Hinsdale in October 1945 (shown here). The Cub would not hit mass production until two years later. *Wisconsin Historical Society / 60321*

FIRST CUB PRODUCTION

Mr. W. A. Salzmann, general superintendent, congratulating works manager Mr. J. E. Harris (seated on the tractor) on completion of the first Farmall Cub (serial number 501) at International Harvester's Louisville Works on May 13, 1947. *Wisconsin Historical Society / 24287* markets) and orchards as target markets for the machine. It also stated that large farms would use it for a light second or third machine, and that large estates or municipalities were good markets for the Cub.[14]

The Cub was promoted heavily in 1946 and was introduced as a 1947 model. Note that it wasn't until June 1947 that the Cub was formally in production at IH's newly opened plant in Louisville, Kentucky. IH purchased the plant in early 1946 for $6,700 from the federal government's War Assets Administration (about $100,000 in 2023 dollars).[15] The plant had formerly been used by the Curtiss-Wright Corporation to build airplanes. Production began on June 23, 1947. It was produced

1948 CUB
The Cub used the same integral frame construction as the Model A and B, with the engine, transmission, clutch, and rear differential housings serving as the frame. The rear tread width was adjustable from 40 to 56 inches, and the relatively tall, wide front end and arched rear differential gave generous ground clearance. Touch Control hydraulics were available as a special attachment. *Kinzenbaw Collection / Lee Klancher*

URBAN MARKETS
Farmall Cub on display at Union Station in New York City in a photo dated July 8, 1947. The Cub was part of IH's marketing efforts to reach estate owners and other people who had property to maintain but did not own a farm. *Wisconsin Historical Society / 24564*

as the Farmall Cub until 1965, when the Farmall name was dropped in favor of International. The International Cub was built in various forms until 1979.

From the name to the introduction hoopla, the Cub was aimed squarely at a new and different market. The rest of the line bore letters or more utilitarian names. A Titan or Mogul appeals to the large farm operator; Super M-TA feels powerful and assertive. This was not so with the name, "Cub." The name certainly fits with the concept that the Cub would be marketed to a group of people who previously considered tractors too large and expensive for their farms.

The Cub was introduced with trumpets blaring and flags waving. Cubs appeared in parades; Cubs were put on display in banks and at fairs. Several promotions featured herds of Cub Scouts piled on the little tractors. Promotional banners and displays proclaimed ease of use and affordability. The company was attempting to bring the tractor to a new group of people, and management marketed the machine

1949 FARMALL CUB

The Cub was also marketed to farmers as a second machine useful for small jobs on the property. *J. W. Goodwin Collection / Lee Klancher*

MOBILE ADVERTISING

This is a circa 1947 promotion for the movie, *The Farmer's Daughter*, in which the main character, played by Loretta Young, comes from a farm. She won the 1948 Oscar for Best Actress in a Leading Role for her performance. *Wisconsin Historical Society / 2108*

anywhere the general public (not just farmers) gathered. In 1947, a corporate brochure celebrating the 100th anniversary of the invention of the reaper touted the Cub as a rising star. Production was expected to top 50,000 units per year, and Louisville Works, where the Cub was built, was gearing up to meet that anticipated demand.

International closely tracked the first 10,000 Cubs sold and came up with some fascinating statistics. Over half the Cubs went to vegetable, grain, and livestock farmers, with only 3 percent going to tobacco farmers. If the company had actually built the Cub with tobacco farmers in mind, the sales to them were a dismal failure. If the tractor was intended for small farms, it was a smashing success. About 75 percent of the Cubs went to farms of less than 50 acres. The Cub tended to be the only tractor owned, with 67 percent of the buyers not owning any other tractor. Also, nearly half of the Cub owners weren't even full-time farmers.

International found a winner with the Cub. The introduction was a mess, with the tractor rolled out publicly before the company had production capacity to build it, but that was outweighed by the fact that the machine was the right product at the right time.

LOEWY DEALERSHIP DESIGNS

In the 1930s, companies began to standardize and optimize the look of their retail outlets to increase brand recognition. Gas stations, home appliance stores, department stores, and packing companies had developed distinctive styles, contracting high-profile industrial designers to create the look. IH had brought on industrial designer Raymond Loewy in the mid-1930s to transform the look of its trucks and tractors. Those designs were rolled out to the public in 1938 and 1939.

Loewy also redesigned the company logo, and later recalled the work with his characteristically tart assessment. "In view of the power and prestige of International Harvester, I thought their trademark was frail and amateurish. The firm's executives asked me to show them what I had in mind. I left Chicago for New York and sketched a design on the dining-car menu, and before we passed through Fort Wayne, International Harvester had a new trademark. It was reminiscent of the front end of a tractor and its operator."[16]

The logo first appeared in the IH annual report in 1944, at the top of the summary page.[17] The logo was used in October 1945 on experimental tractors shown at the large demonstration held at the Hinsdale farm, both on the TD-24 and Cub.

PENNSYLVANIA DEALERSHIP

The dealership above is a unique design built by Frank Barlow. The building survived intact and is owned today by the Pennsylvania International Harvester Collector's Club (Chapter 17), which runs it as a museum. Much of the interior features products and equipment from the original dealership. *Lee Klancher*

It began to appear more often on machines and literature in 1946, and by January 1947, the logo was in wide use on the company's tractors and crawlers.

In addition to machine design and corporate branding, IH became interested in applying uniform design to its architecture. On July 28, 1943, IH commissioned Raymond Loewy to develop a look for its buildings. This was to include interior and exterior design of branch locations as well as sales and service centers (dealerships) and factories.[18]

With the dealerships, the objectives of the new design were spelled out in a 1948 proposal.[19] In addition to improving customer recognition, the goal was to make the buildings more useful and efficient for dealers, easier to build, cheaper to maintain, and to reduce payroll by consolidating office functions.

The building design was straightforward: it was to be tall enough to accommodate the machines inside the space, and long and low with lots of glass. The description was simple; the look was modern and elegant. The original branch house proposal included suggested floor layouts for several of the branches including Pittsburgh, Mason City (Iowa), Jacksonville, and Columbus. The plans for the dealerships were shared, and IH dealerships were encouraged to adopt them. In the end, approximately 1,800 dealerships adopted the look.

Note that not all IH buildings were designed by Loewy. In July 1947, IH announced a contract for construction of a new motor truck and service branch in Milwaukee, designed by Milwaukee architects Grassold and Johnson.[20] In August, another new branch house to be built in Omaha was designed by John Latenser and Sons, a local architecture firm.[21]

The preserved International Harvester dealership in Millville, Pennsylvania, is one of the dealerships built using the IH-supplied prototype dealership plans. Designed by Frank Barlow, he combined elements of the Loewy-created IH dealerships—notably the distinctive vertical slab with the logo on the side—with some of his own ideas. He started the work on the opening day of deer season in 1946 and built the dealership with lumber cut from his own mill. The dealership survived intact, and it is owned today by the Pennsylvania International Harvester Collector's Club (Chapter 17), which runs it as a museum. Much of the interior features products and equipment from the original dealership.

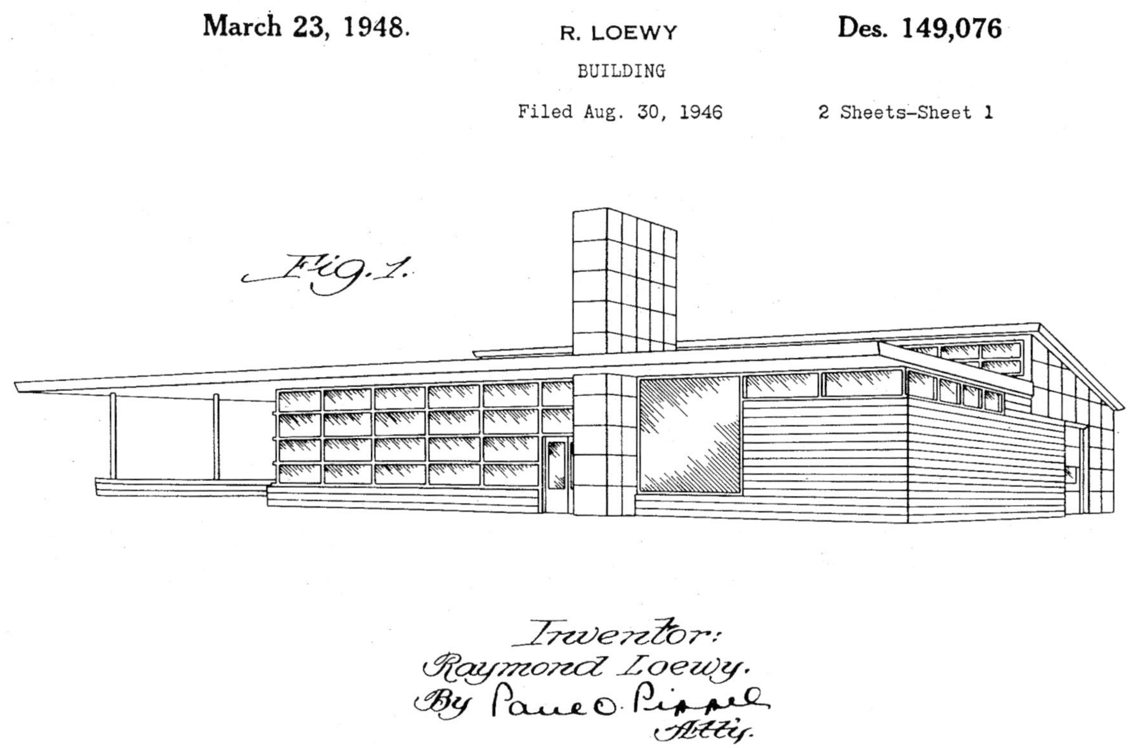

DEALER PATENT

Despite the fact that Loewy's contract with IH stipulated IH owned the patent rights to his creations, in August 1946 Loewy filed for and was granted this patent for the IH dealership design. *US Patent*

BRANCH OFFICE SKETCH

In addition to styling the dealerships, Loewy made style recommendations for the IH branch houses, as shown in this sketch. *Library of Congress*

INTERNATIONAL TD-24

If the Cub was IH's Tyche (the Greek goddess of luck and fortune), the TD-24 was its sword of Damocles. And to clarify—the Greek parable around the sword is that a person is often better off enjoying what they have rather than chasing power and wealth.

In the late 1940s, IH was most certainly chasing new markets. With Fowler McCormick at the helm, the company appeared to foresee that the explosively profitable market for large farm tractors would have to slow down. Fowler's vision for the future led IH into new territory. The TD-24 was part of that exploration.

The TD-24 made an early public appearance alongside the Cub at an October 1945 show held at IH's Hinsdale experimental farm.[22] The big machine was not ready for production. Neither, for that matter, was the Cub, as IH lacked the facilities to produce them. This early showing of not-quite-ready machinery appears to have been a trend at IH in the mid-1940s of pushing to introduce and sell machines before they were ready.

While such actions are understandable—companies are always under pressure to introduce new products and keep sales moving—the practice of rushing heavy equipment to market has had consistently dismal results. Heavy equipment takes years, if not decades, to be properly field tested. If it is not—and nearly every tractor company has at least one rushed-to-production model in its line—the machine winds up being a failure in sales and often a black eye to the company. The John Deere Model GP and 8010 are good examples of rushed development cycles that led to unfortunate models, as is true of the IH 8-16, the flawed final drive on the 560, and the 4100 four-wheel drive.

After the sneak peek at the TD-24 in 1945, the model shows up again in a brief mention in the August-September 1946 issue of *Harvester World*, when the company announced the first production unit (a UD-18 engine) off its newly acquired and retro-fitted Melrose Park Works.

In February 1947, IH hosted an exhibition of industrial equipment at its Melrose Park industrial equipment headquarters and showcased the TD-24.[23] Billed as the "largest and most powerful crawler tractor ever built," IH promised to put the TD-24 into production that year and that it would "set new standards in high-speed handling of big loads."[24] The former statement, at least, would not prove to be true—the Caterpillar D-8 had roughly the same horsepower, and the TD-24 was heavier, meaning the D-8's performance was arguably comparable.[25]

When the machine was sold and put to work, users quickly discovered the tractor's rear end and "Planetary Power Steering" system were not up to handling the power output of the engine. Failures were so common that early TD-24s were recalled and scrapped, and the machine was taken out of production for a few months in late 1948.[26] The TD-24 problems badly tarnished IH's reputation in the heavy equipment field, damaging the company's attempts to compete with Caterpillar and the big Allis-Chalmers crawlers.[27]

The TD-24 came back into production as an improved model in 1949. When tested at Nebraska in 1950, the revised model was too powerful for the dynamometer, so belt horsepower could not be tested. The tractor recorded a stunning 138 drawbar horsepower. Power output rose steadily as the model was developed, with the 1954

INTERNATIONAL TD-24

The TD-24 was a powerful machine designed for construction and large logging outfits. The crawler weighed over 37,000 pounds and was powered by a 1,090-cubic-inch six-cylinder diesel rated for 155 drawbar horsepower. *Wisconsin Historical Society / 10086*

LOGGING IN 1949

The crawler used an eight-speed (forward and reverse) transmission and a 317-pound crankshaft. Despite a host of bells and whistles, a weak transmission in early versions of the TD-24 hurt International's reputation in the heavy crawler market. *Wisconsin Historical Society / 22867*

edition cranking out 155 drawbar horsepower at the Nebraska test, and the 1957 machine topping out at 163 horsepower at the drawbar. Production of the TD-24 ended in 1959.

The crawler weighed about 20 tons and was powered by a 1,090-cubic-inch six-cylinder diesel that carried 28 quarts of oil in its sump. The tractor used an unusual 24-volt electrical system and a 317-pound crankshaft.

The TD-24 was hardly the last of the TracTracTor line, as IH crawlers continued to flow off factory floors in a fairly steady stream. The smaller machines introduced in the 1940s were well-regarded, and the larger TD-14 and TD-18 were popular with GIs who used them during World War II. The TD-24, however, is a piece of the IH puzzle on which too much was spent, production was rushed, and its much-ballyhooed entrance into the industrial equipment market was damaged if not sunk.

PLANETARY POWER

One of the new features on the TD-24 was Planetary Power Steering, which combined hydraulically controlled steering clutches with a two-speed range similar in function to the Torque Amplifier. The two-speed unit was coupled to a four-speed synchromesh transmission, giving eight speeds in forward or reverse. It was not ready for production, and the machine had to be recalled. *Wisconsin Historical Society / 24726*

1949 FARMALL C

The Farmall C was introduced in 1948 as a replacement for the Model B. The tractor was produced in respectable numbers, selling over 20,000 per year in 1949 and 1950. If you include Super C production, more than 170,000 Model Cs were built from 1948 to 1954. *Joe Neville Collection / Lee Klancher*

FARMALL MODEL C

Introduced in 1948, the Model C was a more powerful replacement for the Model B. It was larger and heavier than the Model B, but it used the same narrow front end and was designed to be a two-row cultivator. Like the Model A and B, the Model C was created for the small farmer who was perhaps upgrading from horses to power farming.

The Model C no longer used the dropped-rear drive of the Model A and B. Instead, the Model C used larger rear wheels and a straight axle like those introduced on the F-12 and used on the Model H and M. The straight rear axle allowed the rear tread width to be infinitely adjustable within a specified range (84–100 inches for the M and 44–88 inches for the H), and tread width could be adjusted without taking the rear wheels off. It required larger rear tires but eliminated the extra tooling and complexity of the dropped-rear drive mechanism.

The Farmall C engine was the same 3x4-inch four-cylinder used in the Farmall A and B. By governing the engine to run at 1,650 rpm, 250 rpm higher than its predecessors, a couple of extra horsepower were coaxed from the Model C engine. Burning gasoline, output was about 19 horsepower at the belt pulley and 15 horsepower at the drawbar, enough to pull a two-bottom plow in average conditions with ease. The engine used an IH magneto and a Zenith carburetor.[28]

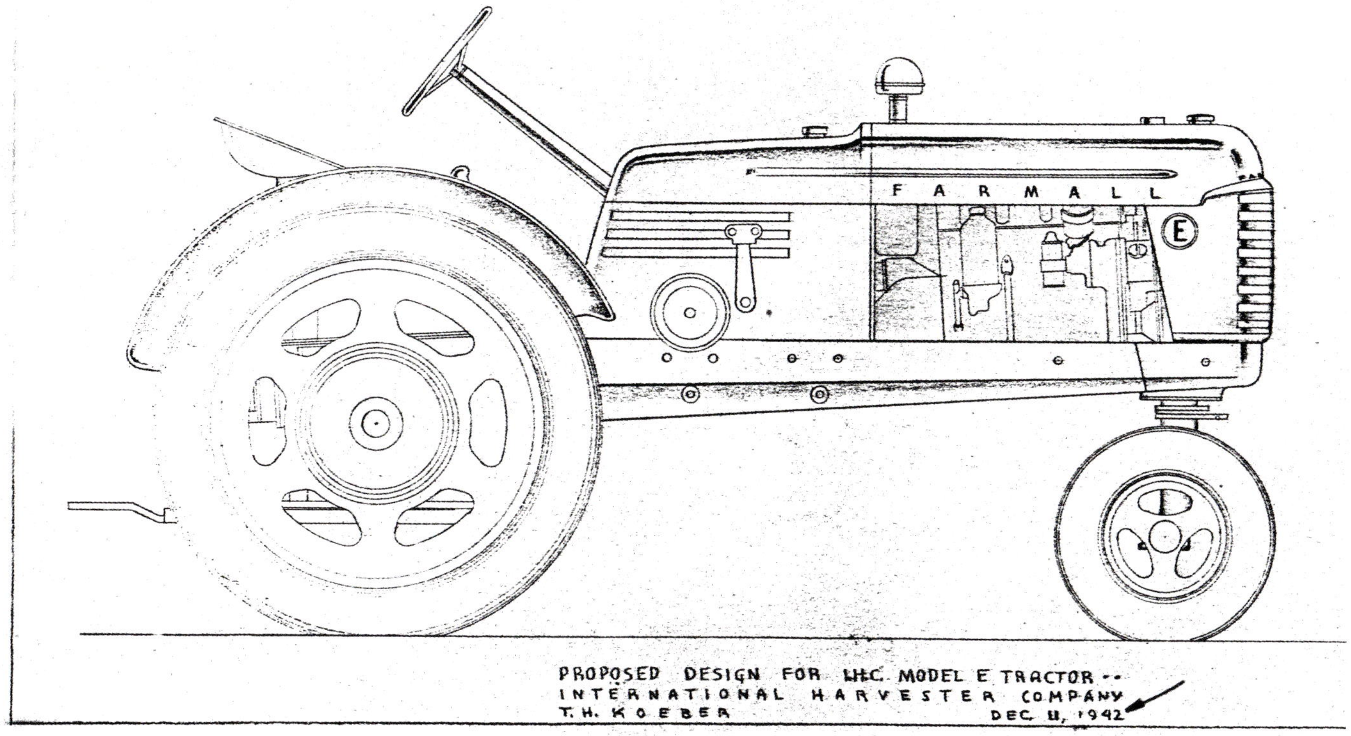

WORKING THE FIELD

The Farmall C used a straight rear axle rather than the dropped-rear drive mechanism found on the Model A and B. The straight axle required taller rear wheels but allowed the rear tread width to be infinitely adjustable within a specified range. It also made adjustment simpler, as the wheels did not have to be removed. *Wisconsin Historical Society / 160875*

MODEL E SKETCH

IH in-house designer Theodore H. Koeber sketched a number of variations, including this Model E tractor from December 11, 1942. *Gregg Montgomery Collection*

1948 FARMALL C WITH ONE-ROW CORN PICKER

The Model C engine was the same C-123 unit used in the Super A, which was simply the engine from the original Model A and B with a 1/8-inch-larger bore. *Brian Hale Collection / Lee Klancher*

TURLEY SPRAYER AND SEEDER

The C. J. Turley Company of Bloomington, Illinois, made this machine, and the caption states it has a sprayer, seeder, and corn detassler attachment. This photo is dated June 2, 1950. *Bill Frank, McLean County Museum of History*

The engine was the same as that of the Model A and B, but the chassis was significantly larger and heavier. The Model C weighed about 3,000 pounds, much heavier than the 1,700 pounds or so of the Model A. The chassis used the engine and four-speed transmission cases as stressed members, like the Model A and B. The front end bolted to the engine cases, and the rear used straight splined axles that allowed the tread width to be adjusted.

The Farmall C also had Touch Control hydraulics, like those on the Super A. Options included a wide tread model, although these are fairly rare. The Model C was quite popular and outsold the Model A and Super A, with peak production at more than 20,000 units per year.

SUPER A/C CHRISTMAS TREE TRACTOR

This unusual machine is one of seven that were custom-built in Canada in the 1950s. The high arch allows tractors to mow grass on Christmas tree farms. Abraczinskas Nurseries of Catawissa, Pennsylvania, purchased three of these mowers in 1953, and two of those mowers were purchased by Frank Bartlow Jr. of Millville, Pennsylvania. The third worked on the nursery until 2019. *Kerns Collection / Lee Klancher*

FARMALL SUPER A

The Super A was an upgraded version of the Model A that featured a more powerful engine mated with a host of optional equipment, including Touch Control hydraulics, quick-attach implements, and the universal frame mounting system.

Touch Control hydraulics debuted on the Super A and the Model C. Touch Control was a hydraulic lift that powered implements up as well as down, giving the operator a powered lift at the touch of a lever. Lifts had been around for quite some time, but hydraulics were an up-and-coming technology. They eliminated the need for clumsy mechanical linkages to a lift and afforded greater mobility to tractor attachments. Unfortunately, they also required horsepower to run. The horsepower diverted to hydraulics made the tractor a bit less efficient at the drawbar. Some even say that a Model A had more pulling power than the Super A due to the horsepower lost to the Super's hydraulic system. Later Touch Control systems had a temperature gauge mounted between the control levers to monitor hydraulic fluid condition. This gauge was used on the Super C.

The universal frame mounting system was another advance designed to simplify the farmer's life. Attaching implements to a tractor's frame typically involved bolting on an attachment frame. Each implement required a unique frame, making changing implements a major pain. With the universal frame mounting system, all IH attachments used the same mounting

FARMALL SUPER A

The Super A was produced from 1948 to 1954. Two distinctly different decals were used on the hood. The earlier decal has white type with a ribbon behind the "A," and later ones have red type on a white circle. The ribbon logo was used earlier in production, and the white circle came later. The common lore is that the Supers should have all received white circles, but IH had some stock of ribbon logos left and stuck 'em on until they were gone. J. L. Aldridge of Waycross, Georgia, is shown planting corn with a Farmall Super A and A-178 planter circa 1952. *Wisconsin Historical Society / 61216* frame, and switching implements became easier. IH's Fast-Hitch actually provided a similar service, as all Fast-Hitch implements used the same mounting system on the tractor.

The International version of the Super A was still known as the International A. It featured all the new Super A bits, plus a beefier Touch Control system with heavier control arms and slightly more powerful internals.

The Super A-1 used a larger bore to bring displacement of the four-cylinder engine up to 123 cubic inches.[29] It also had a thermostat to go with the water pump, bringing the cooling system up to date. The Super A-1 was produced only in 1954. About 2,000 serial numbers were issued that year, so production should have been close to that.

FARMALL SUPER A

The Super A had a four-speed transmission and C-113 four-cylinder engine. This Super A is shown circa 1950. A Super A-1 with a larger engine (the C-123) replaced the Super AV in 1954. *Wisconsin Historical Society / 24727*

FARMALL SUPER AV

The Super AV was designed to cultivate high crops. As with the Super A, the Super AV-1 replaced the Super AV during the 1954 year. *Wisconsin Historical Society / 161177*

MONEY & WIT: THE M&W STORY

Elmo Meiners was a farmer and business owner in central Illinois. While working with his 1941 Farmall M tractor, he came to realize the five-speed transmission was too widely spaced and the operator had to bring the tractor to a dead stop to shift. He took his problem to Art Warsaw, who was a trained mechanic that owned an International dealership in Minier, Illinois, located between Peoria and Bloomington.[30]

In 1946, Warsaw and his friend, Meiners, began working on a solution in the basement of the Meiners Grain Company elevator building in Anchor, Illinois.[31] They tore apart Meiners's M, jammed the drivetrain into the building's elevator, and put it into the basement. They scavenged an electric motor from the building's coal furnace and hooked it up to the M's transmission so they could spin it and test it.[32]

With Warsaw serving as engineer and Meiners as helper, the pair built a high-low range gearset that was installed inside the stock transmission of the Farmall M, turning the five-speed into a nine-speed transmission. It was a high and low range unit that affected only the first four gears. In low, first through fourth were lower and more tightly spaced than factory. In the other setting, the first four gears were higher—but also more tightly spaced. The top transmission speed—fifth gear—was not affected by the M&W kit.

M&W FIRST NINE-SPEED TRANSMISSION

This is the first M&W nine-speed transmission, shown when it was designed in the basement of an office building in Anchor, Illinois. Art Warsaw (center left) and Elmo Meiners (with finger on transmission) disassembled Elmo's 1941 Farmall M and took it into the basement to design the transmission.
Bill Burnham Collection

The M&W nine-speed kit also affected the PTO speed, which required Art and Elmo to design a disconnect to maintain the proper PTO speed. The gears in the M&W kit meshed a bit more smoothly than the stock IH transmission, which meant the M&W nine-speed shifted more quickly and easily. The unit tucked nicely inside the stock transmission case—the entire piece of kit was well-made and clever.

Once the transmission was working in the basement, Warsaw and Meiners hauled it upstairs and reassembled the 1941 Farmall M, a machine they dubbed "Gramps."

They perfected that first transmission on Gramps,[33] and that 1941 M would become their primary test machine.

Bob, the nephew of Elmo Meiners, recalls learning how to drive a tractor on Gramps.[34] He also recalls their test machines looking pretty rough, as they were more concerned about power and performance than shiny paint.

With their nine-speed transmission working, Elmo Meiners and Art Warsaw named their new venture the M&W Company. While one might think the initials were for their last names, the two liked to say it was for "Money" and "Wit," reflecting the fact that Meiners managed mainly promotion and sales (money), and Warsaw did more of the tech development (wit). Installation of the M&W nine-speed kit required disassembly of the transmission, drilling one hole, adding an engagement lever and cable, and dropping in the new gearset. This was fairly straightforward work, but it needed to be done by a skilled mechanic.

Warsaw and Meiners believed the best way to sell their new transmission kit was through IH dealerships. Their IH branch house was in nearby Peoria, Illinois, and it serviced eighty-one dealerships in the region. In fall 1948, when M&W rolled out its new kit for a retail price of $159 ($1,931 in 2023 dollars), those eighty-one dealerships were its target market.

To promote the new nine-speed transmission, the pair traveled around the region, demonstrating it with Gramps.[35]

GRAMPS DEMO

This is the 1941 M after Art and Elmo hauled it out of the basement and reassembled it. The photo was taken outside the office building basement where the nine-speed was designed, Art and Elmo cobbled together the signs you see, and used the Farmall M to demonstrate the new nine-speed transmission at several hundred tractor dealerships in the region. The tractor was later nicknamed, "Gramps." *Bill Burnham Collection*

1940 FARMALL M "GRAMPS"

This tractor was a test and demonstration unit for M&W. The machine was in rough shape at the end of its life, but the owner has it in pristine condition today. *Bill Burnham Collection / Lee Klancher*

GRAMPS SIGNAGE

When Warsaw and Meiners hit the road to sell their first transmissions, they traveled to dealerships in the central Midwest and demonstrated the nine-speed in Gramps. *Bill Burnham Collection / Lee Klancher*

1947 FARMALL M WITH ALL OPTIONS

This Farmall M has every option available from M&W (see list). Owner and historian Bill Burnham is shown in the shop where he restored this tractor and many more.

Bill Burnham Collection / Lee Klancher

They created a paper M&W sign touting the new transmission's capabilities, fastened it to Gramps with masking tape, and hit the road. The pair would demonstrate the nine-speed in Gramps at two hundred dealerships that year.

The men had hit upon a solution to a problem that many farmers were experiencing, and the result was an overwhelming demand for their kit. They built transmissions furiously in the basement of an office building for a local grain elevator company. By early 1949, they had a nine-speed transmission kit built for the H as well as the M, and they were unable to keep up with demand.[36] They purchased an old schoolhouse four miles east of Anchor and hired six employees to build and ship transmission kits.

An interesting piece of lore goes that the transmission kit in the H made it more noticeable that the stock H engine's horsepower was inadequate, and that prompted Art and Elmo to build big-bore piston kits for the Model H to power them up. Bill Burham has heard the story from his father and others, but can't say for sure it is fact!

One of the first new employees was Don Biefeldt, a kid fresh out of high school who sought out Elmo on Christmas Day 1949 to ask him for a job. Elmo hired him on the spot, with Don's first day of work being December 27, 1949. Biefeldt soon became the foreman, and he would stay with M&W until April 1995.[37] The six employees became incredibly efficient, building one thousand kits a month, averaging an amazing fifty kits built each work day.

Don Bielfeldt's relative, Dale Bielfeldt, recalled that Meiners had a gift for listening to his customers and turning their needs into technical improvements. "Elmo had a knack I guess, because he would go out and talk to farmers, stop them in the middle of the field, walk out there and talk to them about what I think we need to do. And he would come home and he would think it out and get the engineers to building it."[38] Dale added that Meiners was always driving around the countryside, talking to farmers.

The next innovation that Warsaw and Meiners tackled came about after Meiners bought an airplane. Elmo Meiners's son, Ron, has fond memories of the company.

"After the nine-speed transmission . . . when my dad lived up in the Watseka area, he joined the Flying Farmers and bought a Piper Cub airplane, and they just held meetings and

1947 FARMALL M WITH ALL OPTIONS

This Farmall M (same as on opposite page) is equipped with all twenty-one M&W options available for the Model M. Note the Super-Snoot Radiator Protector is not shown. *Bill Burnham Collection / Lee Klancher*

ALL THE OPTIONS FOR FARMALL M OFFERED BY M&W

- Add Life Oil Filter
- Add Pow-r Carburetor Kit
- Add Pow-r Dry Type Air Filter
- Add Pow-r Governor
- Clutch Extension (made pedal easier to press)
- Friction Throttle
- Gear Reduction Starter (73 percent more starting power)
- Hi-Lift Rocker Arms
- Hi-Torque Crankshaft
- Jet Black Sleeves (Teflon cylinder sleeves)
- Live Hydraulic System
- Live Power PTO
- Nine-Speed Transmission
- No-Clog Front Wheel Cleaner
- P-61 Dual Hydraulic Control Valve
- PTO Shifter
- Pulley Shifter
- Slip-On Axle Dual Mounts (hubs)
- Super-Snoot Radiator Protector (not shown—put on for harvesting corn)
- Turbo Dome Pistons
- Two-Point Hitch Brackets (to mount Farmall 450 Fast Hitch; very rare—M&W made maybe two dozen of these)

you'd fly in to a breakfast or just fly in to a meeting to talk about farming.

"He and Pete [Elmo's brother] were flying around somewhere looking at farmland or doing something. And Pete wanted to know how come the airplane didn't get hot like the farm haulers did. And dad explained to him that it was air cooled and [had] aluminum pistons.

"And Pete said, 'Well, why can't we put aluminum pistons in our tractor?'"

"Dad said, 'Well, I guess we can!'"[39]

Not only did they create an aluminum piston, they also made the bore larger to increase power and moved the location of the pin to reduce piston slap. The aluminum transferred heat more efficiently than the stock cast iron pistons. On the Farmall M, the M&W piston kit increased the bore from 3⅞ inches to 4 inches and added 9 horsepower. In addition, the M&W pistons came with a one-year warranty; IH replacement pistons were only warrantied for ninety days.

They would also later add a high-domed piston option. Testing of that was done with Gramps—they tore that old engine

1947 EXPERIMENTAL M&W FARMALL H

This H was originally owned by Meiners's brother-in-law and was used as a test bed for the first M&W nine-speed Model H transmission. They showed this tractor at the inaugural M&W show in September 1956, after the new Farmall 350s were released. The M&W folks decided adding white paint to the hood and fenders of the aging Farmall H would make it stand out in the field. The fact that the M&W H produced more than 50 horsepower—10 more than the brand-new Farmall 350—also helped draw attention to the machine. *Bill Burnham Collection / Lee Klancher* apart a dozen times that winter, bolting washers to the top of the pistons, putting it back together, and then putting the tractor on the dyno to see how much horsepower was added.[40] The M&W piston kits were incredibly popular, and Ron Meiners estimated they sold 70,000 kits. He also said customers in the area often requested the dealership install the M&W piston kit (and occasionally an M&W nine-speed as well) in brand new tractors before they left the showroom floor.

"Well, every once in a while you have an International rep go to the International dealership and he goes out in the back to see how things are going and he looks, and here's the seat and everything off of the tractor.

[The IH rep] says, 'What are you doing?'

'Putting in an M&W nine-speed.'

Well, they didn't think that was a good idea."[41]

While IH corporate's initial response to the nine-speed transmission was fairly positive, they weren't as happy about the big-bore kit—particularly when they were getting installed right on the dealership floor.

IH sent letters out to dealerships, alleging that M&W transmission and piston kits put undue strain on the powertrain and caused early wear. M&W folks would counter that their transmission kit improved transmission oiling, and the aluminum pistons had longer life in the field than the stock cast items. IH would also address the shortcomings solved by the M&W innovations, adding more power, improved transmissions, and live PTO to its tractors over the years.

In the meantime, the M&W parts had developed a loyal following and a fan base. Each winter, Meiners would tinker with Gramps, adding parts to add horsepower. Come spring, he would take the beat-up 1941 M out to the field to see how his improvements worked in the real world. His fans knew that and would show up in great numbers to hear the old machine bark and see what Meiners had cooked up for them.[42]

In 1954, Art Warsaw and Elmo Meiners parted ways. Warsaw sold his portion of the business to Bert Ertel of Indianapolis, Indiana. The company continued to grow, moving its headquarters to Gibson City, Illinois, in 1954 and to a larger facility in Gibson City in 1964.

On the afternoon of Sunday, September 27, 1953, M&W hosted its first farm show and displayed its equipment. It placed an ad in the local paper. Much to its shock, roughly five thousand people came to the grounds to see the machines. M&W made the show an annual event and invited local John Deere and IH dealerships to demonstrate their new machines as well. The company held the show for nineteen years—it didn't hold a show in 1972, possibly due to the oil embargo. Its last show was in 1973, when the cost of the show got to be too much (M&W did not charge admission).

Elmo Meiners and M&W continued to listen to farmers in the field, introducing many more innovations over the years. An early live power take-off and rear dual-wheels were designed and sold, as well as high-performance crankshafts, piston sleeves, and improvements to controls, hitches, and more. Meiners retired from M&W in 1989 at age seventy-five. He sold his interests, but remained involved in startups, his community, and the Flying Farmers. He passed away August 25, 2009.[43]

M&W evolved to become RhinoAg, a builder of agricultural mowers, cutters, and other implements based in Gibson City, Illinois. RhinoAg is a division of the Alamo Group, a worldwide conglomerate that owns brands such as Bush Hog, Earthmaster, and many more and is based in Sequin, Texas.

BILL BURNHAM
All the restorations shown on these pages were done by Bill Burnham. He is also the preeminent historian on M&W and has a sizable collection of its history, literature, and historical images. *Bill Burnham Collection / Lee Klancher*

Wisconsin Historical Society / 5248

MID-CENTURY DEMONSTRATORS

For 1950, the IH sales teams unloaded the big guns with the Mid-Century sales campaign, designed to push sales of tractors built in Louisville Works: the Super A, Model C, and the Cub.[44] This marketing campaign included newspaper advertisements, slide films, radio commercials, and direct mail pieces all tuned with the "calibrated gauge of market research."

The core idea of the program was to "try it yourself on this white Farmall . . . the gold-starred, white-painted tractors are making IH the talk of the countrysides. Across the nation, 5,000 machines will comprise the white armada."[45]

In 1950, the IH sales department decided to paint some of the Letter Series tractors white and send them to dealerships as part of a promotional campaign. The tractors were sent to dealerships as demonstrator units, meaning local farmers could borrow the machines for a few days to see if they liked them. IH would boast that more than 360,000 demonstrations took place. The Mid-Century Tractor sales campaign apparently included the W-9 line of tractors.[46] The reference to the W-9 was in a magazine article, and nothing was said about painting the machines white.

Darrel Darst wrote in a 1995 issue of *Harvester Highlights*, the newsletter of the IH club, that white demonstrators were built for three months at the Louisville Plant in Kentucky. According to Darst, the entire line of equipment built at Louisville was painted white.

Records on the white Farmalls are hard to find—as of 2022, no paint committee decision or other official IH paper trail has been found to establish when the first or last one was built, a serial number range of machines created, or much of anything hard. The idea that they were made in the first three months of production—from Darst's article—is accepted lore but not documented fact.

If a farmer purchased one of the white demonstrator models, IH would repaint it red upon request, and many of the original white tractors were documented when owners discovered traces of original white paint under the hood.

1950 FARMALL SUPER A
If a customer bought a white tractor, International repainted the machine red either when it left the dealership or after the white paint became dirty.
Case IH

1950 FARMALL CUB

In 1950, International wanted to celebrate the year with special demonstrators. These models were painted white and loaned out to customers for tests. *Larry Kerns Collection / Lee Klancher*

1950 FARMALL C

The only way to verify an old tractor was a demonstrator is to find the white paint underneath the red, which was the case with this machine owned by Fred Vergin of Loretto, Minnesota. *Fred Vergin Collection / Lee Klancher*

FARMALL SUPER C

The Super C was an increased power version of the Model C. The extra power came from slightly more displacement found with a larger bore, up from 3x4 to 3⅛x4 inches. The engines were now rated by displacement rather than bore and stroke, and the four-cylinder gasoline-burning engine displaced 122.7 cubic inches. The horsepower increase was modest, with a gain of a couple horsepower at both the belt pulley and drawbar.[47]

The Super C introduced International's new double-disc brakes. The disc brakes replaced the drum units and were covered and sealed from grit. Fast-Hitch was offered as factory equipment on the Super C in 1953, along with a plug-in adapter that allowed the use of three-point hitch implements. Note that many Super Cs were retrofitted with Fast-Hitch kits, so 1951 or 1952 models with an IH Fast-Hitch may exist.

1951 FARMALL SUPER C

The Super C was built from 1951 to 1954 and received new disc brakes and a larger cylinder bore for more horsepower. Bob Bennett of Poulsbo, Washington, restored this Super C. "I take 'em down to parade rest," Bennett said. "And he does a lot of cussing," his wife added. *Lee Klancher*

1952 FARMALL SUPER C

This 1952 with a mounted corn planter is owned by Florence Kline of Lititz, Pennsylvania. The tractor was purchased new by Florence's father, Lewis Bidler, and worked on his farm until 1996. *Lee Klancher*

1951 FARMALL SUPER C
The Super C option list included a belt pulley, PTO, electric lights and starter, Touch Control hydraulics, and a Fast-Hitch. This particular example is equipped with Fast-Hitch and Touch Control hydraulics. *Guy Spinelli Collection / Lee Klancher*

FARMALL SUPER M

By the 1950s, IH was looking for ways to breathe some new life into the Letter Series. The line was by no means sagging, as 1950 was a record year for the Model M, but the farmer's appetite for power was slowly but surely increasing. The additions of hydraulics and an increasing number of PTO-driven implements and tools were cutting into the amount of power available to drive the wheels, and 35 horsepower didn't stretch as far in 1950 as it did ten years earlier.

The road to more power for the Model M took the predictable route; the Super M engine's bore was enlarged ⅛ inch to 4 inches while the stroke remained 5¼ inches. The result was a 264-cubic-inch engine that ran at a maximum of 1,450 rpm. Other tweaks included an improved cylinder head, exhaust manifold, crankshaft, and carburetor tuning.

The engine was tested at Nebraska in 1952 and put out 33 drawbar horsepower and 44 horsepower on the belt pulley, an increase of over 8 horsepower. The gasoline engine produced 363 foot-pounds of torque at 991 rpm.[48]

The diesel version was based on the same 264-cubic-inch engine. Like the gasoline engine, it produced peak horsepower at 1,450 rpm. The Super MD engine used a 12-volt system and produced 350 foot-pounds of torque at 1,080 rpm. The diesel machine put out 33 drawbar horsepower and 47 belt pulley horsepower.

The Super M was also tested on LPG gas. The engine's compression was bumped up to 6.75:1 to burn LPG, and the result was even more horsepower and torque. The Super M produced 371 foot-pounds of torque and 40 drawbar horsepower when burning LPG.

1953 FARMALL SUPER M
By the 1950s, IH was looking for ways to breathe some new life into the Letter Series. The line was by no means sagging, as 1950 was a record year for the Model M, but the farmer's appetite for power was slowly but surely increasing. This Super M was purchased new by Willard Holtan on May 15, 1954 and has stayed in the family ever since. *Allan Holtan Collection / Lee Klancher*

1952 FARMALL SUPER M-LP

Wilbur Goehring spotted this tractor at the 1952 South Dakota State Fair and promptly went to Kiel Implement in Highmore, South Dakota, and ordered one. He added the factory LP option—this was one of the first models in which IH would factory equip the machine with propane (earlier LP models were conversions rather than factory order models). This Super M-LP has been on the Goehring farm in Hyde County ever since. The engine still has the original main and rod bearings. The tractor was restored in 1999. *Lee Klancher*

The Super M was released with a plan for four different variants, known as the Stage I, II, III, and IV models. Stage I was the original Super M, which had the larger-bore C-264 engine described earlier and an updated version of the Lift-All hydraulic system from the Farmall M.[49]

Stage II Super Ms were fitted with new live hydraulics. That system required significant changes, including the new hydraulic system as well as a revised clutch housing and the seat from the Super C.

Stage III was the Super M-TA, and Stage IV became the Farmall 400 (both of which are discussed later in this book).

HIGH-HORSEPOWER SUPER M

M&W continued to offer high-performance options to improve the factory offerings from IH. This 1954 Super M is equipped with every M&W option available, including M&W pistons, a stroker crankshaft, and high-rise rocker arms. Along with a few more tweaks, this modified Super M is good for 80 horsepower on the dyno. *Bill Burnham Collection / Lee Klancher*

SUPER MD DIESEL ENGINES 1953

By Jim Allen

The UD-6, UD-9, and UD-16 engines were updated in 1953 with bore increases—the UD-6 to 4 inches and the UD-9 and UD-16 to 4½. This bumped the displacement to 264, 350, and 525 cubic inches respectively. With some accompanying tuning and a 100-rpm bump in the redline, the UD-6 delivered a maximum of 54 horsepower and the UD-9 78.5. These engines saw use in the "Super" versions of the MD, WD-6, and WD-9, as well as with power units. They also appeared in the updated tractor lines like the 400 Farmall, produced from 1954 to 1956, and the W400 standard. The final evolution of the UD-6 engine line came in 1956, with an increase in displacement to 281 cubic inches via a bore increase to 4⅛ inches. The TD-9 engine continued on as it was until it was discontinued in 1958.

LIVE HYDRAULICS

By Kenneth Updike

The Farmall H and M tractors used what is termed the "belly pump" hydraulic system during their production. This term described the location of the hydraulic pump, which was placed in the belly of the clutch housing. The pump had an integral oil reservoir with a fill tube/dipstick that exited the unit under the gas tank. The pump was driven from the transmission counter shaft via a splined gear coupler. By pulling

FARMALL SUPER MV
The Super M was a popular machine, with more than 57,000 built from 1952 to 1954. Note production was split between the Louisville and Rock Island IH plants. Only 245 were Super MVs. *Wisconsin Historical Society / 2511*

FARMALL SUPER M WITH TWO-ROW NO-TILL PLANTER
Development was done on a No-Till planter at Purdue University. It was first demonstrated in the spring of 1947. It was used to plant corn in 1951, and the experiments worked well. IH produced roughly fifty of these units in 1954 and sold about twenty-three of them. *Kinzenbaw Collection / Lee Klancher*

back on a wire control rod, the hydraulic system was "live" and ready to run. Whenever the clutch pedal was released, the pump was running. When the pedal was depressed, the pump stopped turning and the hydraulics stopped.

This setup worked great if the tractor was moving, but if the pedal needed to be depressed to change gears or slow down, the hydraulics stopped. Hydraulic systems can be slow and finicky in cold weather, especially when oils were not well developed, so in northern climates during the winter, the hydraulic system might not even be used. IH developed an option to allow the farmer to disconnect the hydraulic system. The company called this option "Seasonal Disconnect," and it was a simple two-jaw clutch with an external engagement control lever. On tractors fitted with this feature, a short lever is found on the underside of the clutch housing. An instructional decal tells the operator which direction the lever must be positioned to engage or disengage the clutch.

After World War II, the hydraulic industry exploded with new technology developed for the war finding commercial use. Better pumps, seals, control valves, and cylinders were all products of the war effort. Advancements in sealing materials (synthetics versus leather or rubber) lead to higher operating system pressures. A higher-pressure system could lift heavier implements with less strain on the engine's horsepower.

IH was quick to capitalize on these new technological advancements. The Farmall Super A, Farmall C, and Farmall

Cub were post-war era tractors that had a new hydraulic system that IH termed "Touch Control." The hydraulic system worked with the operator simply touching (moving) control levers. The hydraulic pumps were direct mounted to the engine and ran whenever the engine did. This is called "live" hydraulics, as the system is live whenever the engine is running. The larger Farmall H and M each had a live hydraulic pump system added to them when their Super Series versions debuted in the 1950s.

M&W offered kits to convert IH tractors to live hydraulics. An IH conversion kit that was often added to the model H and M tractors included a "live" hydraulic pump, allowing the

1953 FARMALL SUPER MD

Diesel engines and farm tractors made fine bedfellows, and the first Farmall MD was built on January 13, 1941. This 1953 Super MD is equipped with dual rear wheels, a factory-available option that offered improved floatation and was particularly useful in sandy, muddy, or other soft soils. *Jared Sauer Collection / Lee Klancher* tractor to have a live hydraulic system like that of the Super Series and later-made Farmalls. This was an easy install, as the pump was fitted between the distributor-magneto and the front driving gear. Unfortunately, the gears in the timing gear train on these engines were not hardened enough for the extra load that the pump created. The failure of these gears was not unheard of. The Super Series Farmall had hardened gears, and IH learned of the failure during field testing of prototypes. The fix was to replace the "soft" gears of the H and M with the harder gears used in the Super Series replacements.

Even today, if an H or M is being converted to a live hydraulic system, the replacement of the timing gears is recommended to avoid pump drive issues.

FARMALL SUPER BMD

New Zealand is the grand melting pot of International Harvester equipment. The small island has a robust farming economy consisting mainly of smaller farms that work the rugged and often mountainous terrain of the islands. This Super BMD works for a living on the farm of Bill Weir near Timaru, New Zealand. *Lee Klancher*

1953 FARMALL SUPER MD

The Super MD came out in 1953, and 5,199 of them were built during their two-year production span. *Lee Klancher*

FARMALL SUPER H

1954 FARMALL SUPER H

The Super H featured a bit more power and more standard features. Introduced in 1953, the tractor was only built for two years. This one wears its original work clothes. *Kinezenbaw Collection / Lee Klancher*

Introduced in 1953, the Farmall Super H offered some new IH line options, and the engine was punched out for a few more cubic inches and about 18 percent more horsepower. The H's bore of 3⅜ inches was increased to 3½ inches in the Super H, bumping displacement to 164 cubic inches and horsepower to 24 drawbar and 31 belt for the gasoline-burner. Torque was up to 267 foot-pounds at 1,046 rpm.[50]

The early Super Hs saw several changes during production, and IH actually planned four different releases referred to internally as Stage I, II, III, and IV models. The earliest Super Hs—referred to as Stage I Super Hs—were released without live hydraulics, with the battery installed under the fuel tank, and with the C-164 engine, which was larger and more powerful than the one in the Farmall H.[51]

The Stage II models received the option to add live hydraulics with the same seat from the Super C. On models with the optional live hydraulics, the battery was relocated under the seat.

These changes took place throughout production. A former IH employee who went by the moniker "Dr. Evil" claimed Stage I Super Hs are serial number 19226 and before, and Stage II are serial number 19227 and after.[52] Another person, D Slater, posted that the last Stage I Super H is serial number 19233, and all Stage II models were 19234 and higher.[53]

Stage III and IV models were almost certainly never built. IH records show the company planned to include the torque

amplifier found on the Super TA models. Guy Fay wrote in a 2004 *Red Power* magazine article that IH built one experimental Model H with a TA installed. Note that several collectors have built Super H-TAs, but these are reconstructions not factory builds.

The updated C-164 engine received aluminum pistons, a beefed-up crankshaft, and a larger carburetor than the C-152 in the Farmall M. The cooling system was thermostatically controlled, oiling was done by pressurized lubrication, and the oil-bath air filter and replaceable oil filter kept the engine clean. The engine was advertised as being equipped with exhaust valve rotators, which were supposed to keep the valves rotating to prevent deposit build-up and extend valve life.

The Super H featured double-disc brakes rather than the drum brakes of the original Model H. The new brakes used a pair of rotating discs in a sealed housings. Lights and an electric starter were standard, as was a PTO, belt pulley, Lift-All, and the new deluxe hydraulic seat.

HIGH-PO SUPER

This 1954 Super H features M&W upgrades, including an M&W nine-speed transmission and a gear reduction starter. M&W did not offer piston kits for the Super H, as the block lacked enough extra material to allow for a larger cylinder. *Bill Burnham Collection / Lee Klancher*

FLAGGED FARMALL

Double-disc brakes were among the new features offered on the Super H, and electric lights and starting became standard equipment. This is a 1954 example. *Jerry Kuster Collection / Lee Klancher*

SUPER STANDARDS

The Six Series went Super in 1952, with engines providing increased power. The horsepower gains were found with an increased bore, up to 4 inches, with the same 5¼-inch stroke. The standard engine became the C-264 gasoline burner, and horsepower was up 8 at the belt (to 41) and 7 horsepower on the drawbar (to 33). Bear in mind that the increases were magnified by the fact that the new engine burned gasoline, which produced 5–10 percent more horsepower than distillate.

The D-264 diesel engine used the same bore and stroke as the gas engine and put out the same amount of horsepower. The diesel equipment was a few hundred pounds heavier, but otherwise the Super W-6 and Super WD-6 were nearly identical.[54]

Note that the Super W-4 and Super W-6 received some of the Stage I and Stage II improvements applied to the Super M and H, but they are not as well documented as the changes to the row-crop machines.

In 1953, the Four Series went Super, and each tractor was labeled with "Super" on the decal on the front of the hood. Under the hood, the new C-164 engine was up to 164 cubic inches by increasing the bore ⅜ of an inch. Horsepower took a corresponding jump, up to about 30 horsepower at the belt when burning distillate and 33 on gasoline. The I-4 was not released with a "Super" designation, but it received some of the updates applied to the Super W-4.[55]

The Super Nine tractors used an increased bore to bring displacement up to 350 cubic inches. The four-cylinder engine

McCORMICK SUPER W-4
The Super W-4 was not produced in huge numbers, with about 3,000 of the small tractors built in 1954. Equipped with live hydraulics, as some of them were, it is a useful little machine on a small or hobby farm, even today. The 164-cubic-inch engine was good for 33 horsepower on the belt, and the tractor weighed in at just under 4,000 pounds in factory-fresh trim.
Harlan Johnson Collection / Lee Klancher

had giant pistons, with a 4½-inch bore and 5½-inch stroke. The engine could be equipped to burn gasoline, diesel, kerosene, or distillate. The pistons were aluminum, and the cylinder sleeves were replaceable. The diesel engine used fuel injection and had a compression ratio of 15.6:1. The powerful engine put out 530 foot-pounds of torque at 1,039 rpm. The Super Nines were produced in 1953. Essentially the same tractor was produced until 1958 as the 600 and the 650, which were simply Super Nines with slightly different standard and optional equipment.

The Nine Series provided something key to what the farmer of the day wanted and needed: more horsepower. Although horsepower was not yet the deciding factor in a purchase, it would be one of the most important considerations from the late 1950s to modern times.

McCORMICK SUPER W-6

The C-264 engine and a five-speed transmission powered this model, shown running an International hay baler in 1954. *Wisconsin Historical Society / 161009*

McCORMICK SUPER WD-6

This Super WD-6 is harvesting grain with an International Model 64 combine. The tractor was built at IH's plant at Rock Island, Illinois, from 1952 to 1954. *Wisconsin Historical Society / 25232*

FAST-HITCH

International's proprietary hitch system, dubbed Fast-Hitch, was its answer to the Ford-Ferguson three-point hitch. The development most likely began on the "Frameall" tractors, an effort by IH to counter Ford's new hitch.[56] Ferguson's system was effective and popular. After Ford and Ferguson split up, the three-point became available to any manufacturer, but it wasn't available for free. The long tradition of IH using in-house equipment—carburetors, magnetos, and so on—was not about to be dropped to use a feature that was developed on the Ford.

Fast-Hitch was introduced in the first half of 1953 at a show held under a big circus tent at the experimental IH farm in Hinsdale, Illinois, as part of a rollout of seventy-seven new (or improved) pieces of IH equipment. The star of the show was the new hitch demonstrated on Farmall Super C tractors as part of the "Parade of Progress" promotion in 1953.[57] The tractors put on a square dance display, showcasing how quickly and easily users could connect their machines to implements using Fast-Hitch.[58] These square dances were repeated at state fairs around the country, including in Minnesota and North Dakota.

The display was clever and effective, as Fast-Hitch's advantage over the Ferguson system was its ease of mounting. The operator could simply back the tractor into the implement and go, rather than compete in the wrestling match involved in attaching an implement to Ferguson's three-point system. The drawbacks (for the end user) were that Fast-Hitch implements had to be purchased to take advantage of the new system, and the implements could not be used on other brands of tractors.

IH engineers also developed Fast-Hitch to solve the Ferguson system's problem of erratic draft. The plow wouldn't come right out of the ground, but it would vary in depth as soil conditions changed.

In 1945, Gordon L. Hershman came to IH from the military, where he had worked on hydraulics for firing rockets from ships in World War II. He ended up working on Touch Control hydraulics and then developed a hitch to compete with the Ferguson system. He explained that the engineering group felt that the Ford-Ferguson hitch provided erratic draft control, which meant that fields were worked unevenly and inefficiently. "If you were plowing a field and got into heavy going, it came out of the ground. If you were trying to plow 8 inches deep, and you were in different soils in a field, the damn thing would be 8 inches sometimes, and other times it'd be 4 inches. We didn't think that was much good. That's why we came up with the Fast-Hitch."[59]

Besides the advantages of improved draft control, Fast-Hitch was also much easier to hook up. Engineer Robert Oliver also worked on developing hitches, and he recalls that Fast-Hitch was well received because of easy access. "The tools were getting bigger and bigger, and it was much easier to hitch up than to pound those links on with your heel, or [muscle] those tools around to get 'em lined up. With Fast-Hitch, you could just back in, pick up, and go. It was very popular."[60]

DEALERSHIP DISPLAY

IH's answer to the Ferguson three-point hitch was the Fast-Hitch. Once the patents ran out on the Ferguson hitch, IH's clever but proprietary system was knocked out by the more widely available three-point hitch. *Wisconsin Historical Society / 24071*

FAST-HITCH TRACTOR PROMOTIONAL EVENTS
IH's promotional wunderkinds cooked up a tractor square dance with implements being dropped and attached during the performance as a way to showcase how quickly Fast-Hitch worked. The tractors are shown at such an event but not yet actually doing the "square dance." *Wisconsin Historical Society / 11240*

FAST-HITCH ON FARMALL 200
Johnson Collection / Lee Klancher

Despite the advantages of Fast-Hitch, the buyer still preferred the three-point hitch, so IH looked to develop one. The company looked closely at existing three-point hitch designs and built its own system with an eye toward avoiding the weaknesses of the Ferguson hitch.

The improved hitch featured Traction Control, which allowed the operator to vary the amount of ground pressure on the rear tires with four different settings. The system was mechanical and automatically matched ground pressure to load in any of the four settings. The problem of erratic draft control was solved. "It could go over the terrace and the tool never left the ground. It had the best regulation curve in the industry," Oliver said. Fast-Hitch was slowly phased out as the three-point hitch became an industry standard.

1954 SUPER M-TA

The last and best Model M was built in 1954. The Super M-TA line introduced the tractoring world to the wonders of a fully independent PTO and torque amplification. *Timothy Rexford Collection / Lee Klancher*

SUPER TA TRACTORS

The Super M-TA was a sign of things to come for IH. It featured the company's latest innovations—Torque Amplification (TA) and an independent power take-off (IPTO). As was discussed in more detail in the Super M section, these tractors were part of IH's Stage I–IV plan for upgrading the H and M. In the IH Master Plan for M&H to Rule the World, the Super M-TA machines were their Stage III models.

TA was a high-low range that could be shifted on the fly. The drive was a planetary gear unit located between the engine clutch and transmission. Engaging it reduced tractor speed by one-third, giving the operator more pulling power to work with. The addition of TA was the beginning of increased flexibility in tractor transmissions, as manufacturers strove to give the farmer enough gear selections to match power output perfectly to load. The closer the match, the more quickly and efficiently the tractor could perform a task.

IPTO was a simple concept that was complex to apply. The idea was to make the PTO engagement independent of clutch engagement. Previous PTO drives ran off the transmission, which was the most convenient place to find power in the rear

1954 SUPER M-TA PROPANE

IH records on how many LP versions of the Super M-TA were built are sketchy at best. The tanks were fitted by outside companies and some mounted the tanks lengthwise. The crosswise LP tank could indicate this is a rare example. *Ray Gosnell Collection / Lee Klancher*

1954 SUPER M-TA

The Super M-TA was a tractor for the future. It featured independent power take-off (IPTO), which kept the PTO spinning independently of the clutch. Torque Amplification (TA) was an International two-speed unit that could be shifted on the fly, allowing the operator to kick down the final drive ratio a bit for more power when needed. *Kinzenbaw Collection / Lee Klancher*

of the tractor. An IPTO system was engaged entirely separately from the clutch, using a lever. To do this, power had to be transmitted directly to the rear of the tractor, bypassing the clutch. This was accomplished by driving the PTO directly from the flywheel on the Super M-TA.

The Super M-TA was produced in 1954 only and was the forerunner to the Hundred Series. A diesel version was also available and was known as the Super M-TA Diesel. Note that IH documentation indicates Super MD-TA is incorrect.

SUPER M-TA MYSTERIES AND MYTHS

By Kenneth Updike

The Super M-TA tractor model replaced the Super M in both gas and diesel versions. The new feature of the Super M-TA Series was IH's Torque Amplifier. This item allowed the operator to change tractor speeds on the go (without engaging the clutch) by moving the Torque Amplifier lever forward or backward. The Torque Amplifier lever was on the lefthand side of the operator platform near the seat.

People often don't know that a Super M-TA could be purchased with the Torque Amplifier omitted. A Super M-TA

1954 SUPER M-TA HIGH CLEARANCE

Accounts vary, but most state fewer than one hundred Super M-TA HIgh Clearance models were built (fifty-five and sixty-four are both listed by collectors as the number produced). They are rare birds, and very few survived. Taller spindles and rear drop boxes raised the machine to work high crops. *Kinzenbaw Collection / Lee Klancher*

M-TA DIESEL HIGH CLEARANCE

IH records show only twenty-two Super MDV-TAs were manufactured, and enthusiasts estimate that fewer than ten exist in the US. Two MDV-TAs were sold in Mecum auctions. The one shown here is serial number 76856, and it was sold at a Mecum auction held in Walworth, Wisconsin, on August 8 and 9, 2014. Note that the decal placement is not correct. *Mecum Auctions*

tractor that has a Torque Amplifier installed, has a short wing decal that is incorporated into the Super M-TA circle decal. On this wing decal the words "Torque Amplifier" are printed.

If a Super M-TA tractor was built without a Torque Amplifier (it was not uncommon to find these as the Torque Amplifier was an *added cost* option) the tractor had a round decal that was worded Super M-TA and it did not have the wing decal with "Torque Amplifier" added.

A good number of Super M-TA Series tractors sold in the southern US did not have the TA option. For many farmers, the added cost of the TA was not worth it. For others, the extra maintenance and risk of downtime from a TA failure was a factor in the decision to eliminate the TA feature.

It is possible to determine if a tractor is a Super M-TA with no TA. Super M-TA Series have a unique clutch-TA housing

1954 SUPER M-TA DIESEL

The Super M-TA Diesel was produced in 1954 only, with 2,705 built and sold. The decal on the cowl is correct on this restored example. *Jared Sauer Collection / Lee Klancher* that is longer than the one used on the Super M Series tractors. The Super M-TA tractors that were not fitted with a TA simply had a dummy shaft installed that replaced the TA gear set. The non-TA tractors did not have a shorter housing. Non-TA tractors with an independent PTO unit on the rear are a very rare combination.

An interesting operational note is that if the Super M-TA tractor is fitted with a TA *and* a belt pulley attachment, the speed of the belt pulley can be shifted on the go by moving the TA lever. It makes no difference if the tractor is moving or not. What this speed reduction may accomplish is unknown to this author other than it allows the operator to engage a belt pulley driven implement at slow speed to start it and then shift up to full speed for work.

The standard sixes also were given the TA addition in 1954, creating the Super W6-TA and Super W6-TA Diesel. Like the Super M-TA, these two tractors incorporated torque amplification and live PTO. Produced only in 1954, these were a sign of things to come for IH. Only about three thousand or fewer of these tractors were built, making them extremely collectible.

1954 SUPER W6-TA

Production of the standard tread Super TA was just more than 3,000 units in 1954. The Super W6-TA weighed more than 5,000 pounds and was powered by IH's C-264 engine and a five-speed transmission. *Tim Lare Collection / Lee Klancher*

1954 SUPER W6-TA DIESEL

The venerable D-264 diesel powered this model, which was rated for 47.5 PTO horsepower. Just over 3,000 serial numbers were issued for the Super W6-TAs and Super W6-TA Diesels, indicating production was roughly that amount. Note that the decal on this model is incorrect—it should be Super W6-TA with "diesel" on the hood (see Super M-TA Diesel on bottom of opposite page). *Kinzenbaw Collection / Lee Klancher*

BUILDING POWER
Bill Burnham Collection

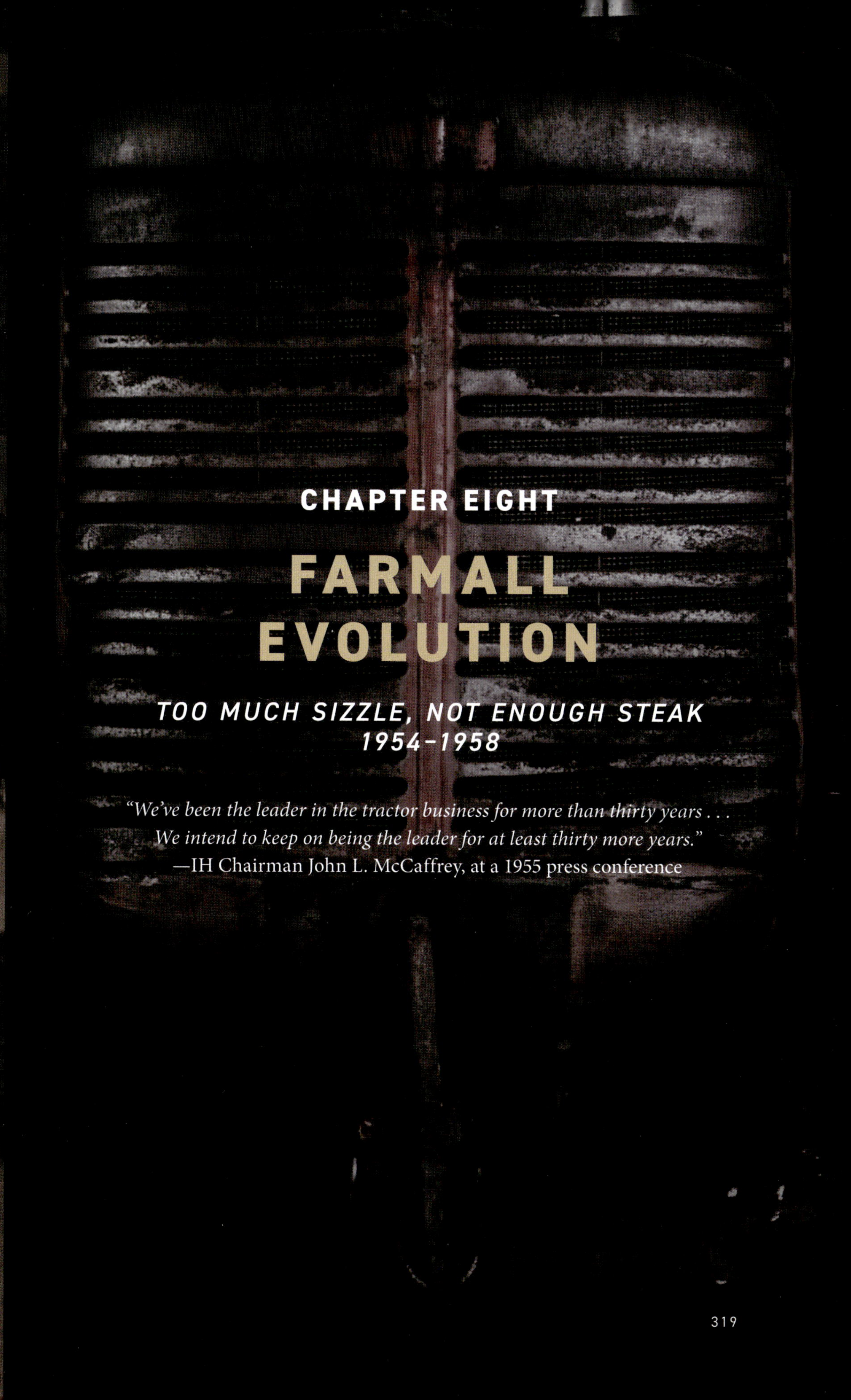

CHAPTER EIGHT

FARMALL EVOLUTION

TOO MUCH SIZZLE, NOT ENOUGH STEAK
1954–1958

"We've been the leader in the tractor business for more than thirty years . . . We intend to keep on being the leader for at least thirty more years."
—IH Chairman John L. McCaffrey, at a 1955 press conference

1950s POWER FARMING

Farmall tractors were widespread on farms across America by the 1950s. Many of those early Letter Series machines had modifications by aftermarket companies like M&W or homegrown improvements that adapted them to the needs of the farmer of the day. *Library of Congress*

While William E. Burnham was off serving in the US Army during the Korean War, his mind was on his farm near Morris, Illinois. His father, Edwin, was running the farm while he was away. William had married Donna Tasdal shortly before leaving to serve, and they had a young son, Bill. William was willing to serve his time—he was also eager to care for his family by working the ground his grandfather had broken.[1]

While William was away in 1956, a used tractor was purchased for the farm, a 1950 Farmall M. His good friend and neighbor, Harold Nelson, was a farmer-dealer for the M&W Company, which built parts that improved farm tractors—particularly Farmalls.

Nelson had a Farmall H. He heard Elmo Meiners and Art Warsaw of M&W were building an overdrive gear for the H, so he sought them out to find out what they had in mind. They offered to install a prototype of the nine-speed transmission they were building for the H in Nelson's tractor. He took them up on that.

They also offered him a job of sorts, selling M&W parts. Nelson was a farmer. He liked the products and Elmo and thought that would be just fine. He was the first farmer-dealer for M&W.

One of the key upgrades M&W offered was a nine-speed transmission that replaced the M's stock five-speed. The additional gears made matching power to load easier in the field and allowed shifting while on the move (you had to stop the M to shift the stock five-speed transmission). While William was away, Nelson installed an M&W nine-speed in the used Farmall M before William came home to the Burnham farm.

The war had its own role to play in the manufacturing world. IH landed several lucrative contracts building 5-ton trucks, Garand M1 rifles, and various airplane parts for the US government during the war. In 1953 alone, the contracts were worth $223 million, helping IH set record sales.

William returned from the Korean War in summer 1956, and the mildly hot-rodded Farmall M proved to be a good workhorse for his farm. The M&W updates—including the

piston kit and other mods—meant it worked nearly as well as the new Farmall 400 introduced that year—at a considerably lower cost.

He needed to be smart with his nickel. In a good year, he and Donna would clear about $3,000 in profits (approximately $37,000 in 2023 dollars). This was about 30 percent higher than the average farm income in 1950, and 10 percent less than the average American family income.[2] It was enough for William and Donna to write a check for $20 in groceries when needed and cover other essentials, maybe buy their son a toy tractor at Christmas—but not much more.

While the Burnhams were managing their farm profitably, IH was struggling. The company had a history of low profit margins, and the mid-1950s were no exception. After dividends were paid, IH took home only 1–3 percent from 1954 to 1958. In 1954, they recorded a 0 percent profit margin.[3]

The workload had been reduced on the American farm, as hours required to harvest crops declined. Starting in the 1940s, new machinery made farmers able to harvest more in less time. Combines dramatically reduced wheat and corn harvesting time, declining from around 100 hours for 100 bushels in 1940 to less than half that in 1960.[4] Milking machines and parlors resulted in even more dramatic reductions in milk production time, dropping the time required to gather 10 cwt (a cwt is 100 pounds of milk, typically about 12 gallons) from 100 hours in 1930 to about ten hours in 1960.[5]

The improvements in equipment didn't mean William had it easy on the farm. He raised a variety of crops and animals on his 300 acres—he had hogs, cattle, and chicken. Just caring for the animals required several hours of work each day.

The Burnham family raised crops as well, but the model was different back then. The corn, for example, was stored in the family's 72-foot-long 8,000-bushel corn crib. Some of it was shelled, and some was ground to feed the animals. Using the corn as feed was more economical than selling it on the market and made the farmer a bit less dependent on market fluctuations. They could always butcher a hog or a cow to fill the freezer.

Nelson and William became good friends. Nelson would periodically make the hour-long drive to have coffee with Elmo Meiners in Anchor, Illinois, and find out what new M&W products he had to offer. Before William left to serve in

NEW HUNDRED SERIES

IH's sales organization was at its peak prowess in the mid-1950s. It showcased its latest at farm shows across the country, including at the 1955 National Plowing Contest held on the Driscoll family farm near Wabash, Indiana, on September 14–17.[i] *Wisconsin Historical Society / 24976* the war, the two spotted an ad in the local newspaper. M&W was holding its first-ever demonstration on Sunday, September 27, 1953, and would have M&W-equipped machinery available for local farmers to test.[6]

Nelsen and William decided this would be a show they couldn't miss and planned to attend. They saw Gramps at work in the field, and M&W had new flat-top oversized piston kits to show off. The show attracted thousands of people—more than anyone anticipated—and Art and Elmo made their show an annual event.

By the time William came back from the war in 1956, the M&W show had become a huge event. The working demonstrations included models of new machines available to test from thirty different local dealers. The stock machines would be on the field alongside the M&W machines, with both available for farmers to drive and see how they stacked up.

M&W SHOW

Elmo Meiners (left) and twenty-seven year old Orion Samuelson at one of the M&W shows, circa 1961. They were broadcasting live on WGN Radio out of Chicago. While the parts M&W built were useful and well-made, the founders were also high-profile showmen. Their tractor demonstration show, held during the fall from 1956 to 1973, drew thousands of people who came to see the M&W-equipped machines go head-to-head with the newest tractor models. *Bill Burnham Collection*

The M&W test tractor, Gramps, was out in the field showing off its hot-rodded powerplant, nine-speed transmission, and a host of other M&W goodies. There was a tractor pull to watch, a tractor economy test, free airplane rides, and free lunch.

Warsaw and Meiners of M&W had a bit to be concerned about at the 1956 show—IH had introduced another new line of tractors—the Farmall 350 and 450. By this time, the factory IH tractors had significant upgrades from the M and H. The 1950s IH tractors featured a larger-bore engine with more power, a two-speed Torque Amplifier that made the M&W transmission kit unnecessary, and improved hydraulics.

M&W painted one of its demonstrator tractors, an H, with white trim to make it look like the new IH machines. It needn't have worried too much. While IH had upgraded over the years, the new-for-1956 IH machines were only minor upgrades over the 300 and 400. That isn't to say IH wasn't good at sales. In fact, the 1950s IH sales group may have been the best in history at moving product, recording all-time high sales figures in 1951 ($1.28 million) and 1959 ($1.36 million) despite a "new" line of machines that didn't offer much more than a new paint job.[7]

The M&W show was a big success, and Nelsen and William weren't the only ones who were impressed—more than five thousand people showed up for the event. It was so successful, Warsaw and Meiners decided to hold it annually. The show became one of the can't-miss farming events of the year.

The show also went on at IH, with its mid-1950s machines and programs, including some useful upgrades, such as live hydraulics, Fast-Hitch, and the Torque Amplifier. The "new" Farmalls of the mid-1950s, however, were too focused on new sheet metal and paint. The chassis underneath was all-too-similar to the Farmall M and H introduced in 1939.

One can also point to the company's heavy investments in refrigeration and construction as costly distractions from its core business of agricultural machinery and its strong business building trucks. One wonders even about the focus on things like Electrall, which were high-technology gambles that didn't pay off.

The refrigeration plant and equipment was sold off to the Whirlpool-Seeger Corporation in a deal struck on September

27, 1954. The Evansville plant and equipment were sold for $19 million, with an estimated $2.5 million in costs for termination pay for the employees. The reason McCaffrey gave was that they needed to either invest heavily in the division, or sell.[8] He didn't mention the sales network, but that certainly appears to be the major issue. People didn't buy appliances at truck and tractor dealerships.

With twenty-twenty hindsight, one can point to the way that Warsaw and Meiners listened to farmers and created solutions to their problems as something IH could have done better in that era. While both M&W and International Harvester knew how to create sizzle, the smaller firm did a markedly better job at engineering solutions that improved the working farmer's day-to-day life.

Thankfully for the legion of farmers with IH red paint in their veins, International's engineering team was turning field data and farmer feedback into higher-horsepower six-cylinder machines that would appear as 1959 models.

SPACE AGE SALES

Led by sales executive John McCaffrey, IH's promotional efforts in the 1950s were over-the-top. The hyperbole in the annual reports is unmatched in this era. While the attention to the sales network did generate results, the engineering suffered from a lack of attention to the basic concerns of the farmer of the day. They needed more horsepower to work more ground. *Wisconsin Historical Society / 24779*

FARMALL 100

Introduced as part of a newly styled line on December 15, 1954, the 100 incorporated the features of the Super A-1, including a one-prong Fast-Hitch, stylish new grille, and, for the model ID, hood and nameplates rather than decals. Touch Control hydraulics were standard on the 100 and 130. This showroom may have been at IH headquarters in Chicago, Illinois. *Wisconsin Historical Society / 58847*

HUNDRED SERIES

The Hundred Series represented further refinement of the Farmall platform released in 1939, with upgraded styling, a little more horsepower, and a host of new equipment. The chassis of each came from the Letter Series introduced fifteen years earlier, with new sheet metal that gave the Hundred Series a cleaner look. The entire new line of Hundred Series was introduced on December 15, 1954, with a total of five different sizes and sixteen models.

The Hundreds also represented complete integration of utility, wheel, and cultivating tractors. The Farmall tractors were the cultivators, while the International tractors were the utility and standard-tread tractors. The cultivators were the standard taller-tricycle or wide-front tractors, while the utility and wheel tractors were lower to the ground with an adjustable, wide-front axle.

INTERNATIONAL 100

This industrial version of the 100 was not produced in large numbers and sold mainly to municipal and commercial customers. Typical industrial options included a foot accelerator, heavy-duty front axle, industrial tires, and an option to remove Touch Control live hydraulics. Customers could also simply order "International" badge plates and install them in place of the "Farmall" plate on the sides of the hood. *Andy Kraushaar*

The first group of Hundreds appeared as the 100, 200, and so on, but IH went through a phase where the name was changed every third or fourth year. The "new" machines received additional numbers (130, 240, 350, etc.), and the plethora of models made the line a bit confusing. Nevertheless, each "new" model resulted in a sales spike, which seemed to be the driving force behind renaming the tractors each year.

Some of the significant features of the Hundred Series had debuted on the later Letter Series tractors: Touch Control hydraulics, independent power take-off (IPTO), and Fast-Hitch (IH's two-point hitch). Two of these three—the hydraulics and IPTO—were improvements that would last. Fast-Hitch would eventually be written out of the play by the three-point hitch, simply because the Ferguson three-point system became the standard for the industry.

"Live" or independent PTO simply referred to a system that kept the PTO shaft turning when the clutch was disengaged. Previously, the PTO did not turn until the clutch was let out. The advantages are obvious—an implement keeps running even when the tractor stops—but the application was not that simple. Power had to be transmitted directly from the engine to the PTO, bypassing the clutch and transmission. To do so required more housings and more moving parts capable of wearing out, but the results were well worth the effort.

FARMALL AND IH 100 AND 130

The 100 Series were basically Farmall Super As with a few new features, slightly more power, and new badges. The tractor was built as the 100 from 1954 to 1956 and as the 130 from 1956 to 1958. Although the machine was refined and restyled along the way, the chassis and engine design was not significantly changed during its amazing production span of forty years.

The Farmall and International 100 engine was the venerable C-123 unit used in the Super C and Super A-1. It was equipped to burn gasoline, which was the standard by that time, but could be equipped to burn kerosene or distillate.[9] The 100's engine produced 18 belt and 14 drawbar horsepower. The Farmall 130 engine put out a little more power, at 20 belt and 16 drawbar horsepower. Although horsepower kept climbing incrementally over the years, the same bore and stroke was used.

Power was transmitted with a nine-inch dry-disc clutch, and the transmission was a four-speed. The same replaceable cylinder sleeves and ball and roller bearings were used as well. The rear wheel tread was adjustable from 40 to 68 inches, and the optional front was adjustable from 44 to 64 inches. The tractor weighed about 2,600 pounds.

By 1958, the cooling system used pump circulation and was thermostatically controlled, leaving the Cub as the last in the line to use a thermosyphon system. Earlier tractors (A, B, Super A, Super A-1) used thermosyphon cooling with an optional booster pump.

The 100 Series came from the factory with a bevy of standard equipment that included a new one-point Fast-Hitch, Touch Control hydraulics, fenders, nonadjustable front end, battery ignition, lights, and a muffler. Other options included a belt pulley, PTO, rear light and taillight, exhaust valve rotators, foot accelerator, radiator shutters, wheel weights, and a host of custom implements and drawbar attachments. A Farmall 100 Hi-Clear was available as well, with the same host of options and attachments.

FARMALL 100 HIGH-CLEAR

The high-clear versions of the 100 and 130 were fitted with larger tires and wheels in the front and rear. They could not be fitted with Fast-Hitch. *Mecum Auctions*

FARMALL 130

The 130 had a mildly hot-rodded version of the C-123 engine that was in the 100. The 130's C-123 ran at higher rpm (and produced more power) than the 100 engine, and it also had lightly modified crankcases, sleeves, valves, crankshaft bearings, and a different governor. *Jay Graber Collection / Lee Klancher*

FARMALL 200 AND 230

The 200 Series was the upgraded version of the Super C. The Super C already had Touch Control hydraulics, so the 200 was not really much of a change. The 200 was produced from 1954 to 1956, the 230 was produced from 1956 to 1958, and the 240 was produced from 1958 to 1962. Both Farmall and International 200 Series tractors were produced, the distinction being the Farmall was the cultivating tractor and the International was the utility tractor.

The 200 Series used the same C-123 engine as the 100 Series, but the 200 put out 22 belt and 16 drawbar horsepower. The engine cooling was thermostatically controlled, and it was equipped to use gasoline, with optional kerosene or distillate equipment. Horsepower increased slightly as the model progressed through the years.

The 200s were equipped with IH's new double-disc brakes, which were enclosed and mechanically operated. Other standard equipment included a temperature gauge, lights, electric starting, Touch Control hydraulics, Fast-Hitch, and muffler. Rear tread width was adjustable from 48 to 88 inches. Options include the Hydra-Creeper, combination PTO and belt pulley, hydraulic remote control, hour meter, and a high-ratio third gear.

1955 FARMALL 200

The 200 is a Farmall Super C with revised styling and a smattering of improvements. The C-123 engine in the 200 was tuned to produce about 10 percent more horsepower than the C-123 in the 100. *Jim Johnson Collection / Lee Klancher*

HYDRA-CREEPER

One of the more interesting options was the "Hydra-Creeper," which allowed the tractor to crawl along at speeds of 0.25 to 1 mile per hour at full throttle, and was available on the 100, 130, 200, and 230 tractors.[10] This attachment was used for transplanting crops or for pulling equipment when harvesting, all primarily for smaller vegetable, flower, or tobacco farms. The unit consisted of a hydraulic pump that was driven by the PTO. The hydraulics drove a chain and sprocket that were geared extremely low. With Hydra-Creeper engaged, the regular transmission was used, but speeds in each gear ranged from about 0.2 miles per hour in first to 1.1 miles per hour in fourth. Hydra-Creeper could be disengaged at will, allowing normal operation of the transmission. To use the PTO, the unit had to be disconnected.[11]

Touted as ideal for a farm of 80 to 120 acres, the 200 filled the niche of an affordable smaller tractor with enough power to pull two 14-inch plows and perform all the tasks necessary on a smaller farm.

FARMALL 200
With a four-speed transmission, disc brakes, and 24 horsepower at the belt, the 200 was a capable little tractor. The model was fitted with the standard two-point Fast-Hitch and Touch Control live hydraulics. Add a McCormick loader, and you have a fine workhorse. *Case IH*

1957 FARMALL 230
This model offered mild engine modifications that gave it more power than the 200, along with significantly improved hydraulics and hitch. It included the new Traction Control Fast-Hitch with optional Hydra-Touch hydraulics. *Jeff Jacobs Collection / Lee Klancher*

FARMALL AND IH 300

The 300 Series was the descendant of the Model H and the Four Series. Like the rest of the Hundreds, new sheet metal, more power, and more options were the order of the day. These mid-sized tractors were available with a Torque Amplifier (TA) and live or independent PTO (IPTO).

The 300 was the result of an evolution that began with the Super Series tractors, particularly the four-stage plan laid out by IH. As discussed in the Super H section, the Stage I and II versions were rolled out during Super H production. The difference between the two was the addition of live hydraulics. Stage III would have added the optional torque amplifier, and Stage IV had even more improvements. The Stage IV Super H was essentially the plan IH used to create the 300.

Note that this was a highly unusual time for IH, with "all-new" models rolled out in rapid succession, each heavier on style than features. This trend was popular with automotive manufacturers at the time—think of the 1950s Chevrolet sedans with their increasingly gaudy fins. This also makes sense for a company led by a salesman—CEO John McCaffrey came from the sales side.

Even the language in the company annual reports became more sensational. The 1953 annual report cites, "more new products than any time in history," yet the year featured a lot of small updates and not many massive launches. The 1955 annual report has a downright bizarre line in the engineering updates, stating that research and development was being conducted on all types of power, including "combustion, electronics, radioactive isotopes, and nuclear reactors." IH annual reports tended to spin everything positively—as one would expect from such a document—but inflated

FARMALL EVOLUTION
These three sketches show the evolution of the Farmall. IH had terminated the contract with Loewy in 1953, and these designs were done internally. One of these has Theodore H. Koeber's name on it. T. H. Koeber was a long-established IH designer, but in the early 1950s he was working mostly on refrigerators. IH industrial designer Gregg Montgomery, who started with IH much later, is skeptical that all these drawings were done by T. H. Koeber. *Gregg Montgomery Collection*

hyperbole and incorrect use of scientific terms lowers the bar and indicates that the sales teams had excessive influence during these years.

The 300s were built from 1954 to 1956, and the 350 was built from 1956 to 1958. The International 300 Utility was introduced in 1955. An International 330 (but not a Farmall 330) was built in 1957 and 1958. As with the smaller tractors, the Farmalls were the cultivating tractors and the Internationals were the standard-tread utility tractors.

The 300s were powered by the C-169 engine. Horsepower was up from the Super H, with 34 belt and 27 drawbar horsepower when burning gasoline. The engine could be equipped to burn kerosene, distillate, or liquefied petroleum gas (LPG). The 300's C-169 four-cylinder engine had a slightly larger bore than the Super H's C-164 engine.

The transmission on the 300 could be ordered as the standard five-speed, or with Torque Amplification (TA), which provided ten effective forward speeds. This was the same system in the MTA, in which a gear reduction system was used, and the farmer could shift on the fly between two slightly different gear ratios, giving two ranges to each gear. This worked great for plowing in soil of varied consistency, or for any other situation that required a quick, subtle change in speed.

The new TA introduced on the Super TA tractors was part of a key factor in the advance of tractor technology: matching load to speed. Research had shown that tractors work most

1956 FARMALL 300

The 300 was the evolution of the venerable Model H, adding a larger bore C-169 engine, independent PTO, Fast-Hitch, and Hydra-Touch hydraulic system. More than 29,000 300s were produced between 1954 and 1956. *J. W. Goodwin Collection / Lee Klancher*

PROPANE WAYNE 1955 FARMALL 300 LPG

This 300 was restored by Stuart Wakeman of Lawrence, Kansas, for the Delo Tractor Restoration Contest. The tractor made an appearance with former Miss America Katie Stamm in a Case IH clothing advertisement. The tractor is dubbed Propane Wayne, and it still works for a living doing general chores on the Wakeman farm. *Lee Klancher*

FARMALL 300 SUPER HIGH-CROP

This Farmall 300 High-Crop has been fitted with an aftermarket kit that raises the tractor another 12 inches. *Louis Wehrman Collection / Lee Klancher*

effectively when power exerted is matched exactly to what is required. Anyone who has plowed with an older tractor where third gear is too high and second is too low understands this need. More gear choices make it more likely that load can be perfectly matched, and the result is less time in the field and better fuel efficiency.

The M&W Company's nine-speed transmission addressed this issue, and engineers across the agricultural industry were working on solutions. While IH's TA was an effective and useful solution, the top tier of development is a powershift or fully hydrostatic transmission, which allows the operator to shift across a full range of the gears while under power. These solutions were expensive to develop. IH would take a very long time to build them for their full-size tractors.

The IH Torque Amplifier was an available option on the 300 models. This could be paired with the optional independent power take-off (IPTO).

The hydraulic system had multiple options, including ordering tractors with no hydraulics. Hydra-Touch controls were available with one, two, or three valves, and hydraulics could be paired with the new weight-transfer Fast-Hitch.

That more sophisticated Fast-Hitch used linkages to improve weight transfer, and was first made available for the 300s in July 1956. It could be retrofitted to previous 300 Series tractors. Power steering became an option in May 1956.

Standard equipment on the 300 included electric starting and lights, available either as factory equipment or as an option to retrofit on models without the options from the factory. The models with carbureted gasoline and distillate engines had six-volt electrical systems, while the diesel and LPG models came equipped with twelve-volt electrical systems. Sealed beam headlights and a tail-light were available for six- or twelve-volt systems.

INTERNATIONAL 300 UTILITY

This variant of the 300 Series was the most significant new model in the line, replacing the Super W-4 with a considerable upgrade that featured an optional Torque Amplifier and independent PTO. Engines equipped to burn diesel, gasoline, kerosene/distillate, and LPG fuel were offered for the International 300 Utility. According to a March 25, 1955, article in the *Chicago Tribune*, the tractor was first displayed in March 1955. *Wisconsin Historical Society / 24632*

FARMALL 300

This 300 is shaking off the dew on a chilly morning in Minnesota. *Don Thune Collection / Lee Klancher*

CONCEPT DRAWING

When developing the 300 Utility, IH was looking for a competitor to the Ford N-Series machines. While this sketch departs considerably from the machine it released, the timing and basic configuration of the tractor appear to be part of the effort to combat the Ford. *Gregg Montgomery Collection*

NATIONAL PLOWING CONTEST APPEARANCE

The three millionth IH tractor produced was a Farmall 300. In September 1955, IH sent its new machine to a national plowing contest held in Wabash, Indiana. Richard Nixon was campaigning there when this photograph was taken on September 17. *Wisconsin Historical Society / 2057*

FARMALL AND IH 350

In addition to the new paint scheme, the new-for-1956 350s featured a weight-transfer Fast-Hitch and push-button starting. The gas engine had a larger bore and more power. They made room for the larger bore by using thin-wall cylinder sleeves.

A diesel option was offered in this line for the first time. IH opted to use a Continental GD-193 engine, available in both row-crop and Wheatland tractors. The International 350 Wheatland model was available in both gas and diesel-powered versions.

Cooling was the standard thermostatically controlled water pump system, and the engine was lubricated by an oil pump. The tractor weighed in between 4,000 and 5,000 pounds, depending on equipment.

When the 350 was launched in 1956, it was touted as an "entire new line of Farmall tractors," when it in fact improved on the previous models with little more than fresh white paint on the hood and grille.

The 300 Series sold well for IH, selling about 35,000 in 1955 alone with more than 100,000 sold from the introduction in 1954 to the close of production in 1963. Like the Farmall H that preceded it, the 300 Series were venerable, useful machines that were used on a wide range of farms and applications.

FARMALL 350 AND BALER

The 350 received the new-for-1956 two-tone paint scheme and the C-175 engine. This one is shown working with a McCormick baler circa 1957. *Wisconsin Historical Society / 8554*

1957 FARMALL 350

This 350 is equipped with M&W upgrades including turbo-dome pistons, an upgraded hydraulic pump, and the M&W governor. The pistons added 1.5 horsepower, and the governor made the engine respond better under heavy loads. The IH engineering team had refined the 350 to the point that it was hard for the M&W guys to find areas to improve. *Bill Burnham Collection / Lee Klancher*

INTERNATIONAL 350 WHEATLAND SPECIAL DIESEL

In 1957, IH offered a Continental GD-193 four-cylinder diesel engine in the IH and Farmall 350s. The engines were direct start, meaning they fired up on diesel (unlike the IH diesel of the time). The Wheatland Special package could be applied to the Utility or the Farmall 350, and it included a different front bolster, an increased clearance front axle, and a different platform with the full-coverage fenders. Wheatland Specials did not use Fast-Hitch. *Lee Klancher*

1957 INTERNATIONAL 350 UTILITY

The 300 Utility was upgraded to the 350 Utility, receiving the updates done to the rest of the 350 line. The old five-speed transmission from the Model H was used in the 350 Series. This one is an IH 350 Utility with a gasoline engine, torque amplifier, and a Fast-Hitch.
Lee Zylstra Collection / Lee Klancher

1957 INTERNATIONAL 350 HI-UTILITY

The Hi-Utility version of the 350 Utility featured larger rear wheels and a raised front axle for increased ground clearance. The machine was made in diesel, gasoline, and LPG versions. This tractor has had an easy life, as it spent thirty-five years doing odd jobs on 40 acres in Minnesota and then living with a collector. It is equipped with Fast-Hitch, power steering, and original hydraulic controls.
Alden Holtan Collection / Lee Klancher

INTERNATIONAL UTILITY 350 HIGH-CROP LPG

Utility models with very high clearance like this one are quite rare. This is also one of only a handful of LPG models built. *Kinzenbaw Collection / Lee Klancher*

INTERNATIONAL 330

Built in 1958 only, this economy model matches the C-135 engine later used in the Model 340 with the 350 Utility chassis. These models were only branded International; no Farmall branding was used with the 330. IPTO and TA were optional; the weight transfer Fast-Hitch was standard equipment. *Jerry Mez Collection / Lee Klancher*

FARMALL 400

The Farmall 400 was an upgrade to the Super M-TA, adding more powerful features, improved hydraulic options, and an upgraded Fast-Hitch. With the same great features as the M-TA but with more models built, this is a great value for those who work the ground with their old iron. This Farmall 400 is shown pulling a disc harrow circa 1955. *Wisconsin Historical Society / 24623*

FARMALL AND IH 400

The successor to the Farmall M and Six Series was the 400 Series. These tractors had a bit more power than the Super M and featured the IPTO and TA that debuted on the Super MTA. The 400 had the new sheet metal and more power as well. They also offered engines that burned LPG, gasoline, kerosene-distillate, or diesel fuels.[12]

The 400 was the result of an evolution that began with the Super Series tractors, particularly the four-stage plan laid out by IH. As discussed in the Super H section, the Stage I and II versions were rolled out during Super H production, the difference being the addition of live hydraulics. Stage III was the Super M-TA, and Stage IV had more even improvements. The Stage IV Super M is essentially the model 400.

The Farmall 400 and W-400 had power steering as an option. The steering systems on the Farmalls and Internationals were quite different, as were the clutches. Both had a five-speed transmission, which was an evolved version of the original five-speed offered on the H & M in 1939.

The 400 and 450 had disc brakes as standard equipment. The discs were changed several times during production.

The 400 was the only model to be offered with Electrall, at least when it was introduced in December 1954. You could put Electrall on other tractors—see Electrall section for more details on this interesting option. The International W-400 used the chassis from the W6-TA and offered power steering. The 400 Series four-cylinder engine was the C-264 from the Super M, although it had more snort, with the standard gasoline burner putting out 45 belt and 36 drawbar horsepower.

FARMALL 400 WITH TREE DIGGER AND TRACKS

This 400 is shown working in Brookdale-Kingsway Nurseries in Bowmanville, Ontario, in October 1956. The custom-constructed machine had Arps tracks and custom shielding and was used to extract trees out of the ground. *Archival & Special Collections, University of Guelph, Rural Heritage Collection, XA1 RHC A0449*

FARMALL 400 DIESEL

The diesel 400s and 450s had twelve-volt systems, upgraded from the six-volt systems on the gasoline models. IH offered a kit to upgrade any of the 400 or 450 six-volt systems to twelve volts. *Mecum Auctions*

FARMALL 400 LPG

The engine for the LPG models was a modified version of the gasoline engine. The pistons, sleeves, and cylinder head were changed to give the LPG engine a higher compression ratio. The 400's 24-inch diameter fuel tanks were changed to 22-inch fuel tanks on June 22, 1955. This 1958 Farmall 400 LPG is pulling a Model 21 cotton stripper near Lubbock, Texas. *Case IH*

1955 FARMALL 400 HIGH-CLEARANCE

The high-clearance 400s were available with gasoline, diesel, and LPG engines. This one was fitted with Char-Lynn power steering. *Mecum Auctions*

INTERNATIONAL W-400

The W-400 was heavily based on the Super W6-TA it replaced. The model was available with diesel, gasoline, and LPG engines. They were offered with operator platforms and full-coverage rear fenders. *Andy Kraushaar*

McCORMICK 120 COTTON PICKER ON FARMALL 400

The 300 Series and 400 Series tractors could be used on cotton pickers such as this McCormick 120. The pickers were designed to be removed and the tractor could be used for normal farm tasks. *Wisconsin Historical Society / 24624*

1957 FARMALL 450

The "all-new" 450 offered mid-century farmers, who were looking to upgrade from a 400, a larger displacement engine, two-tone paint, and not a whole lot more. Under all that pretty paint, fancy hydraulics, and upgraded power train you'll find the bones of a 1939 Farmall M. *J. W. Goodwin Collection / Lee Klancher*

FARMALL AND IH 450

The 450 offered the new white and red paint scheme, C-281 and D-281 increased-power engines, weight-transfer Fast-Hitch, and push-button starting.

Gas, distillate/kerosene, diesel, and LPG engines were offered, and the Farmall 450 was available in high-crop versions. The engines had a larger displacement than the 264-cubic-inch four-cylinder in the 300 due to an increased bore and thinner cylinder sleeves. The 281 gas engine also differed from the 264 in that it had a new carburetor, revised manifold, and a new battery ignition.

The 450 gas models were thirsty animals and used quite a bit of fuel, which pushed the sales of LPG models in the day. As time passed, diesel engines would become much more popular than LPG models.[13]

The 400 and 450 gasoline models offered two different fuel tanks, one with an 18 gallon capacity and a larger 21.5 gallon tank.

FARMALL 450 LPG

The use of LPG gas in farm tractors peaked in the 1950s. As the decade faded, diesel overtook LPG as a popular alternative to gasoline or kerosene. *Marv Rohlena Collection / Lee Klancher*

1958 FARMALL 450 DIESEL HIGH-CLEARANCE

The 450 high-clearance models were built with gasoline, diesel, and LPG engines. Fewer than thirty of these diesel high-clear 450s were produced. *Kinzenbaw Collection / Lee Klancher*

1957 INTERNATIONAL W-450

The IH version of the 450 was popular for heavy field use, so it was widespread in parts of the country with large, open plots of ground to work—here's looking at you, Great Plains. *Mecum Auctions*

INTERNATIONAL W-450 LPG

This model is equipped with optional wheel weights. *Andy Kraushaar*

THE 450 AND THE SUPER-SNOOT

M&W's new ideas often came from farmers, as one of the founders, Elmo Meiners, was fond of driving around the countryside and stopping in to talk with his customers. The Meiners family farmed as well, so some of the company innovations emerged due to family experience. Bob Meiners is the nephew of Elmo. Bob spent much of his childhood working on the experimental farm, seeing new products being built or tested on the dynamometer.

He remembers working with a brand-new Farmall 450 and one of the first corn picker/shellers sold by IH with his brother. "Glen and I decided we were going to go out and pick up corn behind the picker. And we got our gunny sacks and we about froze to death and we started collecting the corn. And we thought a bushel of corn was my grandmother's metal bushel basket, so we started filling the bushel basket and we had about 20 of them in piles going through the barnyard. And Elmo drove out then, and dad and him were talking something about that picker sheller. They were getting heated about it."[14]

The tractor was overheating, and Elmo and his brother, Peter, were concerned the block would crack. Both were upset at the heavy loss of corn from the new device. They figured out that a snoot would keep the chaff out of the engine. They unbolted a snoot they had on Peter's John Deere combine and sent it down to the M&W shop. The guys modified the snoot to mount on the 450—it was ready to work the next day.

Bob and his brother headed back into the field. At the end of the day, they approached their father and uncle, looking for a check.

"That's not even a bushel of corn," Peter said reproachfully, eyeing the thin pickings from their day's work.

"Just pay them," Elmo said.

Elmo's own payday would come when M&W introduced the Super-Snoot. The device protected the engine from dust and dirt and shielded farmers from dangerous fires. Farmers appreciated the innovation, and the Super-Snoot sold well.

1957 FARMALL 450

The M&W Company continued to offer tweaks and upgrades that improved IH machines. This one has a stroker M&W crankshaft, a custom grind on the camshaft, and other modifications that boost the stock 55 belt horsepower to 75. *Bill Burnham Collection / Lee Klancher*

SUPER-SNOOT BROCHURE

This is the M&W Super-Snoot, developed to keep the air intake clean when harvesting corn. *Lee Klancher Collection*

ELECTRALL

By Jim Allen

In the early 1950s, International Harvester was innovative and unafraid to market new things. One of those things was the "Electrall," which debuted on June 22, 1954, when IH and GE engineers showed off an experimental version to press and other guests at the Hinsdale Farm. IH President John McCaffrey spoke at the event, lauding the new innovation as a "the sort of forward-looking engineering which we like to emphasize."[15] Electrall was put into production in early 1955[16] in the form of a tractor-mounted generator kit that cranked out 10 kilowatt three-phase 120/208 volts.

IH developed the technology in conjunction with engineers from General Electric to give the farmer a more versatile power source in the field. "The possibilities of this are far-reaching," McCaffrey said. "It may make possible very desirable changes in the fundamental design of important farm machines." IH stated that the technology could power combines and balers, power portable farm tools, provide portable lighting, and power the farm during outages.

Even in the early 1950s, the electrification of rural farms was not complete, nor was it particularly reliable in many places. As a result, many farmers had standby generators. They weren't there just to keep the house lights on and the fridge going, either. This was back in the days when many farmers did it all. A few dairy cows, steers, hogs, sheep, chickens, plus crops. Farmers with a small dairy had electric milking machines and refrigerated milk tanks, and anyone who knows the dairy business knows a power outage can ruin the farmer's day . . . and old Bessie's.

On top of that, International Harvester had some ideas about the way implements should run. Shaft-driven PTO implements were just getting a foothold in the tractor market and only the newer tractors had them. At that time, most implements like hay balers and pull-behind combines were self-powered using

1957 FARMALL 450 WITH ELECTRALL

One of the big initiatives for IH in the 1950s was Electrall, which was introduced with great fanfare at a demonstration at the IH experimental farm at Hinsdale in June 1954. Company president John McCaffrey spoke, saying, "Its potential is so great, that we thought it desirable to make a progress report on our work even though research is not finished." This nicely finished example is serial number 909 and is equipped with fenders, lights, a fuel gauge in the fuel cap, Fast-Hitch, PTO, front weights, and Electrall. *Huber Collection / Lee Klancher*

ELECTRALL BROCHURE

Snazzy graphics and exciting features were not enough to make more than a small handful of people purchase an Electrall. The new technology quickly faded to obscurity.

Case IH

ELECTRALL ON 1957 450

The Electrall option could be fitted to the 400 and 450 tractors as well as the Super M-TA and W-TA tractors. The McCormick 55 baler had an electric option designed to be powered by Electrall.

Lee Klancher

small gasoline engines. In the case of the baler, the engine ran the baling machinery, and the tractor towed the baler and hay wagon. IH contended self-powered implements (including the McCormick 55 Baler)[17] were more efficiently powered by an electric motor fed juice by a tractor-mounted generator. It could probably have also been argued that an electrically driven implement offered a more precise and constant speed, less mechanical complexity, better efficiency, and more safety over a PTO-powered unit. IH offered kits that could attach Electrall directly to the Super M-TA, Super W-6-TA,[18] 400, and 450 model tractors that were equipped with independent PTOs or IPTO Torque Amplifiers.[19]

Unfortunately, the idea failed miserably. In typically quirky fashion, IH had shot holes in its own idea by introducing the Electrall just as it was introducing live PTO in its tractors, and live PTO was to become the farmer's best friend. So, IH adapted and began to offer PTO-driven generators mounted on a trailer or on a Fast-Hitch-mounted skid. By then, the electric implement idea had gone completely by the wayside, but IH still had a lot of generators left to sell. This idea brought more acceptance and IH soon upped the PTO generator rating to 12.5 KW.

The final variant of the Electrall debuted in 1956 for the S-Line light trucks. It was the same 12.5 KW unit then offered for the tractors, but it was mounted in the bed of a truck and driven by a transmission PTO. It could be installed on a new truck or retrofitted to an older one. The truck units were potentially useful in a commercial or agricultural environment, but they didn't generate any hoopla and were quietly discontinued early in 1957. The various tractor units were discontinued not long after and today they are very rare and seldom seen, even by IH enthusiasts.

FARMALL 450 ENGINE SWAPS

By Jim Allen

GUMZ FARMS TRACTORS
The late Galen Wilkinson (left) and Louis Wehrman (right) ended up with the two Gumz Farms tractors and are seen here at the 2013 Red Power Roundup. Note the engine in Wilkinson's tractor is painted the incorrect color. Wehrman's tractor (in the background) shows the engine painted the correct "Cummins Old Gold" color. *Jim Allen Collection*

In the late 1950s, Cummins Engine Company did some trolling to test the diesel repower market for tractors. According to Farmall collectors Galen Wilkinson and Louis Wehrman, Cummins invited the owners of twenty-eight larger Indiana farming operations to its factory in Columbus, Indiana, to sell the idea of repowering tractors with Cummins engines. One operation, Gumz Farms of North Judson, Indiana, stepped up to buy two kits. Rather than repowering two used tractors, Gumz went full bore by ordering two new Farmall 450s from IH without engines. One was ordered with a wide front axle and the other retained the standard narrow front. They were received in December 1958, which was right at the end of Hundred Series production. By serial numbers, the unit with the wide front is a late '57 build and the narrow front was made well into '58 production. Both tractors then had new Cummins J4-70 four-cylinder diesels installed at the Gumz farm shop with input and advice from Cummins. They were up and running by the middle of 1959 and worked on the Gumz operation chopping peppermint for ten years until sold. By all reports, these tractors did their work well, but the sales results of the repower kits were underwhelming, with only a few known to be sold. Miraculously, both the Gumz 450s have survived though a succession of owners and were eventually restored to their original splendor.

The Cummins J-Series engines debuted as 401-cubic-inch six-cylinders in 1950, an evolution of the A and AA Series diesels that started production in the late 1930s. They were modern, direct-injection diesels with a 4.13x5-inch bore and stroke that used the Cummins PT injection system. The

FARMALL 450 REPOWER

The rather plain side of the J4-70 engine showing the oil filter built into the oil pan. *Jim Allen Collection*

267-cubic-inch four-cylinder Model J4 appeared in 1957 and shared architecture with the six-cylinders. The engine chosen for the Farmall 450 swap was the lowest rated J4 in the lineup, the J4-70, with the "70" indicating the rated power at 2000 rpm. Rated torque was 198 lbs-ft at 1300 rpm. Next up was the J4-80, which made 80 horsepower at 2500 rpm and 298 lbs-ft at 1400 rpm. A JT4 turbocharged version, also known as the J4-120, was rated at 117 to 120 horsepower at 2500 rpm and 383 lbs-ft of torque. There were many other variants of this engine. The J-Series gradually morphed into the C-Series in 1959, and the last of those were built in the early 1970s.

Compared to the D281 gas-start diesel that a 450D used, the J4-70 had a little more maximum power and about the same peak torque, though with a broader curve. The J4-70 was probably about as much as the 450 powertrain could handle safely long-term. Between what Gumz paid for the tractors ($3,230 each) and the engines ($1,950 each), was it worth it over buying a 450D for a little less money? Difficult to say this far down the road—and only the Gumz's accountant knew for sure—but they served well.

CUMMINS POWER

This is the pump side of the J4-70, showing the Cummins PT (*P* for pressure and *T* for time) injection pump. The PT was one of the better systems at the time and could be called a predecessor to today's common rail injection systems. *Jim Allen Collection*

FARMALL CUB (1954–1958)

The Farmall Cub received a bit of a makeover for 1954. The new radiator grill was the most noticeable feature, with the fuel tank, hood, and decals also changing.

For 1955, IH produced a lowered version of the Cub, the International Cub Lo-Boy. The tractor used a nonadjustable front axle and a 1,640-pound weight. Designed for industrial use, the low-slung little machine was a popular choice for everything from factory floors to orchards. It was produced until 1968, with just under 24,000 built.

The Cub and Cub Lo-Boy received a new two-tone red and white paint scheme for the 1956 models. The engine speed was upped a few rpm to give a tad more power in early 1956.

1954 FARMALL CUB

The Cub received a styling upgrade in 1954. Beginning that year, a one-point Fast-Hitch was available to mount on units that had Touch Control hydraulics. *Kern Collection / Lee Klancher*

INTERNATIONAL CUB LO-BOY

The Cub Lo-Boy was introduced in 1954. It was available both as a Farmall and as an International. Some of the Internationals were painted yellow. *Graber Collection / Lee Klancher*

1955 CUB WITH HIGH AND WIDE

The high and wide attachment was available for the Cub, either from the factory or as a dealer add-on. *Kerns Collection / Lee Klancher*

1955 FRENCH CUB

This Cub was built in 1955 at the IH plant in Saint-Dizier, France. The French Cubs were manufactured almost entirely in France. This example uses a Solex carburetor, which gave the Cub 1.25 more horsepower than the American version. The French Cub was built from 1955 to 1957, when it was replaced with a French Farmall Super Cub. *Mike Schmudlach Collection / Lee Klancher*

1956 FARMALL CUB

In 1956, the Cub received a two-tone paint update to match the larger machines. *Larry Kerns Collection / Lee Klancher*

CUB LO-BOY

The smaller Cub also received the styling treatment. The Cub was one of the most prolific and long-lived IH tractor models, with more than 200,000 built between its introduction in 1946 and the last original design model built in 1964. Variations of the Cub remained in production until 1981. *Lee Klancher*

1960 CUB FORKLIFT

This Cub was converted to a forklift by Piper & Paine of Nunda, New York. *Larry Kerns Collection / Lee Klancher*

INTERNATIONAL 600 AND 650

The Model 600 didn't appear in the IH line until 1956—several years after the rest of the line. Produced for only a few years in the late 1950s, the big 600 and its successor, the 650, were upgraded versions of the Super Nine models.

The 600 and 650 used the big 350-cubic-inch engine from the Super Nines and were rated for 60 belt horsepower. Weighing in at over four tons, the 600 also boasted all IH's latest innovations—IPTO, power steering, and an optional cigar lighter.

The 350-cubic-inch four-cylinder engine had coffee can-sized pistons, with a 4½-inch bore and 5½-inch stroke. The standard engine burned gasoline, but it could be equipped to burn kerosene or distillate. Diesel and LPG versions were available as well. The pistons were aluminum, and the cylinder sleeves were replaceable. The diesel engine used fuel injection and had a compression ratio of 15.6:1. Battery ignition was standard, as were lights and an electric starter.

An International 650 was tested at Nebraska in 1957, and it recorded 55 belt and 43 drawbar horsepower. The horsepower figures are respectable, but the number that best describes the power of the 600 and 650 is the 530 foot-pounds of torque produced by the powerful, low-revving engine. That's as much torque as a big-block V-8 engine.

The test tractor burned gasoline and weighed more than 10,000 pounds with wheel weights. In the low-gear maximum pull test, the 650 snorted over 8,500 pounds of force—more than 80 percent of its weight.

The engine was cooled with a thermostatically controlled radiator and fan system. The five-speed transmission was mated to a 12-inch single-plate clutch. Top gear was for road use and ran the 600 up to over 15 mph. Disc brakes and 14x34-inch rear tires were standard.

Optional equipment included the IPTO, Hydra-Touch hydraulic system, belt pulley, foot-operated throttle, rice field tires, magneto ignition, high-altitude pistons, radiator shutters,

1956 INTERNATIONAL 600 DIESEL SN 501

The 600s didn't come to market until 1956. The engines were the C-350 and D-350. This is the first 600 built, serial number 501, and it is equipped with a D-350. *Andy Kraushaar*

1956 INTERNATIONAL 600

The big 600 was a derivative of the W-9 and, like the W-9, was available with diesel and gas engines. Rated as a five-plow tractor, the 600 featured power steering, double-disc brakes, Hydra-Touch hydraulics, a 350-cubic-inch four-cylinder engine, and a shipping weight of about 6,000 pounds. Interestingly, an independent power take-off or Torque Amplifier were not available on the 600. *Kim Spaulding Collection / Lee Klancher*

1958 INTERNATIONAL 650 LPG
The 650s had engine upgrades that squeezed more power out of the C-350 and D-350 IH engines. *Kinzenbaw Collection / Lee Klancher*

INTERNATIONAL 650
The C-350 engine featured a combustion chamber designed to swirl the fuel mixture. These "high-turbulence" engine designs were popular innovations of the time, and they helped reduce deposit build-up in the combustion chamber. *Andy Kraushaar*

tachometer, hour meter, wheel weights, heavy duty rear axles, and a cigar lighter (farmers who used tractors with over 530 foot-pounds of torque apparently smoked cigars rather than cigarettes).

The 600 was built only in 1956. Diesel production started in April 1956, with serial 501, and ended in November 1956. The gas models were all built in September and October of 1956. Note that a few production lists will show some "1957" model 600s—this would be those last 600 diesels built in November, as IH considers a tractor built in November or December the next year's model.[20]

The 650 diesel began production in November 1956 and there was a 1957 model as well. Gas and LPG models were built in 1957 and 1958. Production of the 650 ended in December 1958.

1970 FARMALL 1456 DEMONSTRATOR

This original condition demonstrator wears the scars of a lifetime of use and includes a large number of interesting options. *Scherb Collection / Lee Klancher*

FARMALL POST-1958

"We know our competition is smart . . . and we know they are shooting at us."
—IH Vice President Mark Keeler, 1958, at introduction of 40 and 60 Series at Hinsdale

This section is a photographic guide to the models that wore the Farmall brand starting in 1958. For in-depth coverage of IH, Case IH, and Steiger tractors, combines, and company history, see the first three books in this series: Red Tractors, Red Combines, *and* Red 4WD Tractors.

As the 1950s faded, IH remained the industry leader, but as Mark Keeler pointed out, IH was well aware the competition was gunning for it. This competition fostered one of the most incredible eras in tractor history.

Sales wunderkind (and engineering's nightmare) John L. McCaffrey had pushed the company to introduce products and innovations too early and had failed to get the dollars required to the engineering department. In 1958, this would haunt him. The flagship of the new-for-1959 line, the 560, was introduced with a rear end that wasn't up to the engine's horsepower, and a massive recall would cost IH millions.

Leadership at IH would change hands, and the black eye from the recall and the looming competition would get the engineering department the resources it needed. The Farmall brand would also live on through this era. The 40 and 60 Series row-crop tractors carried the name on the side of their hoods, as did the exciting new 706 and 806.

The 706 and 806 were the start of a terrific line of IH machines with revamped transmissions and high horsepower—the answers to the farmer of the day's needs. This was quickly followed with more—including the stunning 1206, the first IH row-crop tractor to top 100 horsepower.

In 1971, the next great white hope of International stepped in as company president. Brooks McCormick, the cousin of Fowler and grandson of Cyrus's younger brother William, brought energy and vision that hadn't been seen since Fowler ran the company in the 1940s and 1950s.

Unlike Fowler, Brooks had the dedication to back up his vision, and International made great strides in the first half of the 1970s. Brook's bottom line was fairly simple: choose markets that were likely to be profitable and cull the industries that weren't making money.

The results were impressive, with International making record profits and garnering front-page coverage. In this almost unrivaled era of competition for the farmer's dollar, red and green invested heavily in research and development. Each time International came out with a new, improved tractor, Deere was there with a new machine to up the ante.

Meanwhile, IH's investment in the Farmall brand declined. The badge moved from the hood to a small insignia on the front engine panel. In 1973, the Farmall brand was dropped. The story goes that IH had thousands of badges printed and used them on the tractors until they were gone. When the new 86 line came out, the Farmall badge was dropped from the line.

In 1977, Brooks hired Archie McCardell, a former Xerox CEO who came to IH hailed as a high-tech savior. Under McCardell's leadership, technology would indeed take precedence. The results for the tractor division were revolutionary, with the cobbled-together genius that was the 2+2 articulated tractor debuting in 1979 and the high-technology 88 Series emerging in 1982.

McCardell was not a flawless leader—his ill-advised decision to take on labor led to a massive 1979 strike. A crippling farm recession and a debt load that International couldn't overcome proved to be too much for the company to absorb. On November 26, 1984, International's ag division was purchased by Tenneco, owners of the Case line of tractors.

Under the combined flag of Case IH, the Farmall brand would return to the company's line of utility tractors. While some lament the fact that the flagship machines of the line don't get a Farmall badge, perhaps Bert Benjamin and Alexander Legge would approve. The pair both believed fervently in a small tractor affordable enough for everyday farmers, and that belief drove them to create the original Farmall.

That original Farmall is one of the most important machines in history, one that played a key role in the transformation of our society. Its legacy is cemented in the halls of agricultural history.

FARMALL FANTASY

Concept drawing of a modern-day, high-horsepower Farmall.

Gregg Montgomery Collection

1961 FARMALL 340

The 340 was a new platform housing the C-135 engine coupled to a five-speed transmission and optional TA. *Jim Johnson Collection / Lee Klancher*

FARMALL 140

The 140 was essentially a reskinned 130, using the same C-123 engine and a four-speed transmission. An IH and Hi-Clear version were offered. *Wisconsin Historical Society / 24646*

FARMALL 240

The 240 added Hydra-Touch hydraulics and an optional Tel-a-Depth control system to the 230. An IH utility version was also offered. *Case IH*

1959 FARMALL 460

The new six-cylinder engine was the technical star of the 460, and it was available in gasoline, kerosene/distillate, direct-start diesel, and LPG versions. *Wilson Collection / Lee Klancher*

FARMALL 460 HI-CLEAR LPG

The 460 was offered in this configuration as well as in an IH 460 Utility that replaced the 350 Utility. An orchard model and high utility were also offered. *Huber Collection / Lee Klancher*

1961 FARMALL 560 WITH ELWOOD FWA

IH 560 Wheatlands were also offered. This tarted-up example is equipped with the 560 power-up kit from IH and an Elwood front wheel assist. *Kinzenbaw Colllection / Lee Klancher*

FARMALL 560 HI-CLEAR

The high-clearance variant of the 560 used dropped rear axles and longer front spindles as well as different wheel and tire combinations to gain ground clearance. *Lee Klancher*

1962 FARMALL 560

The new six-cylinder was offered in gasoline, kerosene/distillate, direct-start diesel, and LPG versions. The D-281 diesel engine was good for 59 PTO horsepower.

Max Armstrong Collection / Lee Klancher

FARMALL 504

Available with diesel, gas, and LPG engines and a five-speed transmission mated with TA, the 504 also offered IPTO and double disk brakes. *Lee Klancher*

FARMALL 504 CHISHOLM-RYDER

Based in New York, the Chisholm-Ryder Co. converted farm equipment into specialty harvesters for grapes and other unique crops. *Tim Lare Collection / Lee Klancher*

FARMALL 504 HI-CLEAR

This taller variety of the 504 offered additional clearance for tall crops. The 504 came with a three-point hitch as standard equipment instead of IH's proprietary Fast-Hitch. *Lee Klancher*

1961 FARMALL 404

The 404 was powered by the C-135 engine (gas and LP) and was offered in Farmall and IH utility versions. The late *Harvester Highlights* editor Darrell Darst is at the helm. *Lee Klancher*

1967 FARMALL 706

Equipped with a new sixteen-speed transmission and powerful six-cylinder engines, these tractors were everything the 560 should have been. Diesel engines were originally the D-282, which were later replaced by the 310. *J. W. Goodwin Collection / Lee Klancher*

1964 FARMALL 706 HI-CLEAR LPG

In addition to high-clearance models, the 706 was available with front wheel assist or in Wheatland versions as well as torque-sensing three-point hitches. *Adam Koch Collection / Lee Klancher*

1965 FARMALL 806

Good for 94.5 PTO horsepower (gas engine), the 806 was offered with gas, diesel, and LPG engines. The transmission was an eight-speed manual backed by a two-speed TA. *Don Villwock Collection / Lee Klancher*

1967 FARMALL 806 LPG

The 806 was available in hi-clear, Wheatland, and industrial configurations, with wide and narrow front setups. *Ryan Peters Collection / Lee Klancher*

1965 FARMALL 1206 PROTOTYPE

The 1206 was the first IH row-crop tractor with more than 100 PTO horsepower. This one is serial number 511, a rare bird uncovered by restorer and historian Kristin Gall. The machine is adorned with period-correct demonstrator decals. *Lee Klancher*

1968 FARMALL 756

More power was the name of the game in the mid-1960s, and the 756 delivered 76 horsepower to the PTO when equipped with a gasoline engine and TA. *Huber Collection* / *Lee Klancher*

FARMALL 656 AND 544

The 656 was released in 1965 with gas or LP engines, TA, and more. The 544 came out in 1968, replacing the 504 model. This image is dated October 1969. *Wisconsin Historical Society / 124372*

1964 FARMALL CUB

The Cub received a styling update as well. *Bob Bennett Collection / Lee Klancher*

1969 FARMALL 856

The 856 diesel with TA was capable of 100 PTO horsepower and was available with FWA, hi-clear, and in Wheatland and industrial trim. The Farmall badge was taken off the hood and moved down near the model badge. *J. W. Goodwin Collection / Lee Klancher*

FARMALL 1256 FWA

A TA-equipped diesel model 1256 offered up to 116 PTO horsepower, and row-crop and Wheatland versions were available. The FWA was an option. *Brian Look Collection / Lee Klancher*

1970 FARMALL 1456

With 131.1 PTO horsepower out of the diesel version, the 1456 was a weapon in the horsepower races of the mid-1960s. International Wheatland and Farmall row-crop versions were offered. *Coy Winslow Collection / Lee Klancher*

1971 FARMALL 826

The 826 had an optional hydrostatic transmission. The hydro drained horsepower, making the dual-range manual transmission a popular choice on the 826. *Mez Collection / Lee Klancher*

1970 FARMALL 1026

For 1970, IH offered demonstration models in gold paint. The models included were the 544, 656, 826, 1026, and the 1456. *Lee Klancher*

1971 FARMALL 1026 FWA

Front wheel assist was an increasingly popular option in the 1970s, as the demand for power and traction increased on the farm. This one is made by American Coleman of Littleton, Colorado, a builder of all-wheel-drive trucks who was contracted by IH to build front-wheel-assist units for its 06 Series and later tractors. *Richard Walker Collection / Lee Klancher*

1972 FARMALL 666 SERIAL NUMBER 7500

The last of the 66 Series to be introduced was the 666, which replaced the 656. This is serial number 7500—the first 666 built in 1972. *Jerry Kuster Collection / Lee Klancher*

1972 FARMALL 766

New diesel engines were the star of the show for the 66 Series, offering more power and better fuel efficiency for the new line. The 766 also could be had with a gas engine. *J. W. Goodwin Collection / Lee Klancher*

1971 FARMALL 966 HI-CLEAR

The diesel D-414 in the 966 was good for 96 PTO horsepower with a TA and 91 horsepower with the hydrostatic transmission. *Derek Frey Collection / Lee Klancher*

FARMALL 1066 TURBO FWA

The 1066 added a turbocharger to the D-414, which boosted PTO horsepower to 116. *Chuck Tesh Collection / Lee Klancher*

1972 FARMALL 1066 TURBO

This 1066 is equipped with an aftermarket Hiniker cab—the diamond shape gave it more interior room than other aftermarket options. IH built and sold an optional cab.

Marc Johnson Collection / Lee Klancher

FARMALL 1566 TURBO

The highest-powered row-crop tractor in the line, it offered 161 PTO horsepower from the DT-436 diesel engine and TA transmission combo. *Lee Klancher*

FARMALL 1466 TURBO

The 1466 used the DT-436 diesel engine that produced 133 PTO horsepower. *Mike Wiegman / Lee Klancher*

1974 FARMALL 1468

The V-8 Farmalls look and sound great, but they make less power than their six-cylinder line mates. The IH engine was from the truck line and was adapted for farm use. *Jason Sweeter Collection / Lee Klancher*

FARMALL HYDRO 70 HI-CLEAR

The Hydro 70 was basically a 666 Hydro with different graphics. The same treatment was given to the 966 Hydro, which was rebadged the Hydro 100. *Kinzenbaw Collection / Lee Klancher*

1974 FARMALL 1568

Farmall badging was put on hiatus after the models shown on this page. The 86 and 88 Series lines would not bear the Farmall name. *Joel Jakobs Collection / Lee Klancher*

2008 FARMALL 40

In 2001, the Farmall name reappeared on the scene as branding for Case IH's new line of subcompact tractors, the D and DX models. The line was revised in 2008, with the letter prefixes dropped and badging simplified to just "Farmall" with the model number. *Case IH*

2018 FARMALL 65A

Case IH later added A, B, and C to the model designations, and the line evolved into a wide variety of machines ranging from 35 to 140 horsepower. *Case IH*

2021 FARMALL UTILITY 140A

The largest Farmall models are rated for 140 horsepower. This one is shown pulling a RB465 baler. *Case IH*

2023 FARMALL 120N

The line is available in orchard and narrow ("N") configurations, as shown here. This one has the 100th Anniversary badge. *Case IH*

THE FUTURE OF FARMALL
Case IH introduces its first electric tractor, the Farmall Utility 75C. It brings power, versatility, and operator comfort and includes a portable power source. *Case IH*

EARLY IH TRACTOR DATA 1906–1918

Early IH tractor production is confusing, as models were built in different factories with quickly changing drives and more, so more detail is included on these models.

Model Name	Horsepower*	Drive**	Years	Factory	Number Produced***
Friction-Drive Tractor	10, 12, 15, 20	F/F	1906–10	Upper Sandusky, Akron, Milwaukee	376 (U) 446 (A) 236 (M)
Type A Gear-Drive	12, 15, 20	G/G	1907–11	Upper Sandusky, Akron	248 (U) 356 (A)
Type A Two-Speed, Friction Reverse	12	G/F	1909–14	Upper Sandusky	65
Type A Two-Speed, Gear Reverse	12, 15	G/G	1910–17	Upper Sandusky	203
Type B Gear-Drive	20	G/G	1908–12	Upper Sandusky, Akron	255 (U) 46 (A)
Type B Two-Speed	20	G/F	1910–18	Upper Sandusky	383
Mogul Type C Gear-Drive	20	G/F	1909–14	Milwaukee	2,441
Mogul Type C 25-hp	25	G/F	1911–14	Milwaukee	862
Titan Type D 20-hp	20	G/G	1910–14	Milwaukee	274
Titan Type D 25-hp	25	G/G	1910–14	Milwaukee	1,757
Titan Type D 45-hp	45, 60	G/F	1910–15	Milwaukee	1,319
Titan 30-60	60	G/F	1914–17	Milwaukee	176
Mogul 45-hp (later known as 30-60)	45	G/F	1911–17	Tractor Works	2,437
Mogul Junior 25-hp	25	G/F	1911–13	Tractor Works	812
Mogul 10-20	10	G/F	1912–13	Tractor Works	85
Titan Convertible Road Roller	20, 25	G/G	1912–14	Milwaukee	90
Titan Type D 18-35	35	G/G	1912–15	Milwaukee	259
Titan Type D 30-hp	30	N/A	1912	N/A	1
Mogul 15-30	30	G/F	1913–15	Tractor Works	527
Mogul 12-25	25	G/G	1913–18	Tractor Works	1,543
Steward Tractor	NA	N/A	1913	Deering	10
Waite Tractor	NA	N/A	1913	Deering	2
Mogul 20-40	40	N/A	1914–15	Tractor Works	10
Whitney 6-12	12	N/A	N/A	N/A	0

* Horsepower is rated by IH; it is roughly equivalent to belt horsepower.

** Drive refers to gear or friction-drive, and both forward and reverse are listed. For example, "G / F" indicates a gear forward drive and a friction reverse.

*** Note that machines produced at more than one factory are listed on different lines.

Source: McCormick / IHC Archives, "Tractor Production Schedule," as part of the document, *History of a Tractor,* by A. C. Seyfarth.

Note that some models may not have been produced in all years. For more specific information on tractor data, *Farmall Tractor Data Book* by Guy Fay is highly recommended. For a list of post-1958 models, see *Red Tractors 1958-2022*.

Early McCormick-Deering and IH Standards

International 8-16	1917–22
McCormick-Deering 10-20	1923–39
McCormick-Deering 10-20 Narrow Tread	1926–34
McCormick-Deering 15-30	1921–29
McCormick-Deering 22-36	1929–34
Industrial 20	1923–29
Industrial 30	1931–40
McCormick-Deering W-30	1932–40
McCormick-Deering W-12	1934–38
McCormick-Deering Fairway 12, O-12	1934–39
International I-12	1934–38
McCormick-Deering O-14, W-14, I-14, Fairway 14	1938–39
McCormick-Deering WA-40, WK-40, W-40, WD-40	1935–40
Industrial 40	1936–40

Farmall F-Series Tractors

Farmall (Regular)	1924–32
Farmall F-12	1932–38
Farmall F-14	1938–39
Farmall F-20	1932–39
Farmall F-30	1931–39

Letter Series Tractors

Farmall A	1939–48
Farmall B, BN	1940–48
Farmall C	1948–51
Farmall H, HV	1939–53
Farmall M, MV, MD, MDV	1939–52
Farmall Cub	1947–58
International Cub Lo-Boy	1955–58

Late McCormick-Deering Standards

McCormick-Deering W-4, O-4, OS-4	1940–53
McCormick-Deering W-6, WD-6, OS-6, ODS-6, I-6, ID-6	1940–52
McCormick-Deering W-9, WD-9, WDR-9, WR-9	1940–54
Farmall Super W-4	1953–54
Farmall Super W-6, WD-6	1952–54
Farmall Super W-6TA, WD-6TA	1954
Farmall Super W-9	1953–56
Farmall Super WD-9	1953–56
Farmall Super WDR-9S	1953–56

Farmall Super Series

Farmall Super A	1947–54
Farmall Super A-1	1954
Farmall Super C	1948–51
Farmall Super H	1953–54
Farmall Super M	1952–54
Farmall Super MTA	1954

Farmall 100 Series

Farmall 100, 100 High-Clear	1954–56
Farmall 200	1954–56
Farmall 300	1954–56
International 300 Utility	1955–56
Farmall 400	1954–56
International W-400	1955–56
International 600	1956
Farmall 130, 130 High-Clear	1956–58
Farmall 230	1956–58
International 330 Utility	1956–58
Farmall 350	1956–58
International 350 Utility	1956–58
Farmall 450	1956–58
International W-450	1956–58
International 650	1956–58

CHAPTER 1

1. Forrest Crissy, *Alexander Legge 1866–1933* (Chicago: self-pub., Alexander Legge Memorial Committee, 1936), 44.
2. Crissy, *Alexander Legge 1866–1933*, 48.
3. James Stewart, "The Economics of American Farm Unrest, 1865–1900," *EH.net Encyclopedia*, February 10, 2008.
4. William T. Hutchinson, *Cyrus Hall McCormick Volume 1: Seed Time 1809–1856* (New York: The Century Co., 1930), 17.
5. Hutchinson, *Cyrus Hall McCormick Volume 1*, 15.
6. Interview with Jo Anderson by W. M. Hanna, from McCormick, Cyrus Hall, 1809-1884. Evidence to Prove Cyrus H. McCormick Invented the Reaper, 1831-1931, n.d., housed at Wisconsin Historical Society, McC Mss 8A.
7. Jo Anderson profile by Herbert and Lucile Kellar, September 28, 1955, from McCormick, Cyrus Hall, 1809-1884. Evidence to Prove Cyrus H. McCormick Invented the Reaper, 1831-1931, n.d., housed at Wisconsin Historical Society.
8. Jo Anderson profile by Herbert and Lucile Kellar.
9. Hutchinson, *Cyrus Hall McCormick Volume 1*, 79–81.
10. Hutchinson, *Cyrus Hall McCormick Volume 1*, 83.
11. Interview with Jo Anderson by W. M. Hanna.
12. Hutchinson, *Cyrus Hall McCormick Volume 1*, 86.
13. Karl Rhodes, "Reaping the Benefits of the Reaper," *Econ Focus*, Third/Fourth Quarter 2016.
14. Hutchinson, *Cyrus Hall McCormick Volume 1*, 86.
15. Hutchinson, *Cyrus Hall McCormick Volume 1*, 45.
16. Hutchinson, *Cyrus Hall McCormick Volume 1*, 98.
17. Cyrus McCormick, *Century of the Reaper* (Boston: Mifflin Co., 1931), 27.
18. McCormick, *Century of the Reaper*, 28.
19. Hutchinson, *Cyrus Hall McCormick Volume 1*, 259.
20. Hutchinson, *Cyrus Hall McCormick Volume 1*, 265.
21. McCormick, *Century of the Reaper*, 36.
22. Hutchinson, *Cyrus Hall McCormick Volume 1*, 271.
23. "McCormick-IHC Collection Cyrus Hall McCormick," Wisconsin Historical Society.
24. Elaine Shigley, "McCormick's Patent Woes," *Chicago History Examiner*, November 21, 2013.
25. "Six Degrees of Abraham Lincoln: A Historic Mirror Stored at The Conservation Center," The Conservation Center, April 19, 2020.
26. Doris Kearns Goodwin, *Team of Rivals: The Political Genius of Abraham Lincoln* (New York: Simon & Schuster, 2005).
27. Cynthia Weatherill, "John H. Manny Factory," *Rockford Historical Society Nuggets of History* newsletter, Nov./Dec. 1966.
28. John F. Steward, *The Reaper: A History of the Efforts of Those Who Justly May Be Said to Have Made Bread Cheap* (New York: Greenberg, 1931), 84.
29. Steward, *The Reaper.*
30. Charles W. Marsh, *Recollections, 1837–1910*, Northern Illinois University, Abraham Lincoln historical digitization project.
31. Arthur Pound, foreword to *The Reaper*, by John F. Steward.
32. Vernon Derry, *Plano, Illinois: Birthplace of the Harvester* (Illinois, 1954).
33. Herbert N. Casson, *The Romance of the Reaper* (New York: Doubleday, Page & Company, 1908), 79.
34. Utah State History, "John F. Steward 1841–1915," *Utah History Quarterly* 16–17, nos. 1–4 (1948–1949).
35. Steward, *The Reaper.*
36. Utah State History, "John F. Steward 1841–1915."
37. Casson, *The Romance of the Reaper*, 52.
38. A. W. Harris, *William Deering, born in Maine, 1826, died in Florida 1913* (Chicago: self-pub., 2014).
39. Sam Moore, "The Journey of William Deering," *Farm Collector*, August 1, 2013.
40. Moore, "The Journey of William Deering."
41. Casson, *The Romance of the Reaper*, 53.
42. Casson, *The Romance of the Reaper*, 67.
43. Steward, *The Reaper*, 289.
44. Hutchinson, *Cyrus Hall McCormick Volume 1*, 66.
45. Hutchinson, *Cyrus Hall McCormick Volume 1*, 123.
46. Hutchinson, *Cyrus Hall McCormick Volume 1*, 125.
47. William T. Hutchinson, *Cyrus Hall McCormick Volume 2: Harvest 1856–1884* (New York: Appleton-Century Company, 1935) 42–50.
48. Hutchinson, *Cyrus Hall McCormick Volume 2*, 87.
49. Mary C. Shields to C. H. McCormick, Oct 5, 1865, in McCormick Archives.
50. Hutchinson, *Cyrus Hall McCormick Volume 2*, 131–33.
51. Hutchinson, *Cyrus Hall McCormick Volume 2*, 363.
52. Steward, *The Reaper*, 291.
53. Hutchinson, *Cyrus Hall McCormick Volume 2*, 362–63.
54. Hutchinson, *Cyrus Hall McCormick Volume 2*, 365.
55. Hutchinson, *Cyrus Hall McCormick Volume 2*, 134.
56. Hutchinson, *Cyrus Hall McCormick Volume 2*, 200.
57. "The M'Cormicks," *Chicago Tribune*, October 9, 1872.
58. Steward, *The Reaper*, 335.
59. McCormick, *Century of the Reaper*, 90.
60. Hutchinson, *Cyrus Hall McCormick Volume 2*, 381.
61. McCormick, *Century of Reaper*, 91.
62. Paul Avrich, *The Haymarket Tragedy* (New Jersey: Princeton University Press, 1984).
63. Art Young and John Nicholas Beffel, *Art Young: His Life and Times* (New York: Sheridan House, 1939) 80.
64. Gordon Winder, "A trans-national machine on the world stage: representing McCormick's reaper through world's fairs, 1851–1902," *Journal of Historical Geography* (2007).
65. Ron Chernow, *Titan: The Life of John D. Rockefeller, Sr.* (New York: Vintage Books, 2004).
66. Crissy, *Alexander Legge 1866–1933.*
67. Crissy, *Alexander Legge 1866–1933*, 82–84.
68. Crissy, *Alexander Legge 1866–1933*, 122.
69. McCormick, *Century of the Reaper*, 108–109.
70. Stella Virginia Roderick, *Nettie Fowler McCormick* (New Hampshire, Richard R. Smith Publisher, Inc., 1956), 182.
71. Luther Conant Jr., *The International Harvester Co.* (Washington, DC: US Department of Commerce and Labor Bureau of Corporations, 1913), 58–59.
72. John Garraty, *Right-Hand Man: The life of George W. Perkins* (New York: Harper & Brothers, 1957).
73. Chernow, *Titan*, 375.
74. McCormick, *Century of the Reaper.*
75. Garraty, *Right-Hand Man*, 126.
76. Garraty, *Right-Hand Man*, 134.
77. Roderick, *Nettie Fowler McCormick*, 224, 226.
78. Chernow, *Titan*, 419.
79. William Z. Ripley, *Trusts, Pools and Corporations* (Boston: Ginn and Company, 1915), 331.
80. Ripley, *Trusts, Pools and Corporations*, 331.
81. Ripley, *Trusts, Pools, and Corporations*, 333.
82. *United States v. International Harvester Company*, 1, testimony of William C. Lane.
83. *US v. International*, 1, testimony of W. M. Reay.
84. Ripley, *Trusts, Pools, and Corporations*, 326.

CHAPTER 2

1. "E. A. Johnston Joins the Keystone Company," *The Implement Age*, 23, no. 8 (April 1904): 24.
2. Kenneth Durr and Lee Sullivan, *International Harvester, McCormick, Navistar: Milestones in the Company that Helped Build America* (Oregon: Graphic Arts Center Publishing Company, 2007), 57.
3. Randy Leffingwell, *Farmall: The Red Tractor that Revolutionized Farming* (Massachusetts: Voyageur Press, 2007), 18.
4. Lynn Webster Ellis and Edward Aloysius Rumely, *Power and the Plow* (New York: Doubleday, Page & Co., 1911), 41.
5. George M. Robinson, "The Charter Gasoline Engine," written recollection circa 1925.
6. Lee F. Burger, son of Franz Burger, on IH Chicago office letterhead, to William Robinson, Charter Gas Engine Co., March 14, 1927.
7. Burger to Robinson.
8. Burger to Robinson.
9. A. C. Seyfarth, *Tractor History 1910–1934* (Wisconsin Historical Society, 2008), 18.
10. George Ellis, interview by E. A. Hunger, January 25 and 26, 1935, in Wisconsin Historical Society mss 6z, box 185, folder 17327.
11. "Dolliver's Auto-Mower," *Ottumwa Semi-Weekly Courier*, July 23, 1901, 7; *The Muscatine Journal*, July 23, 1901, 6; *Sioux City Journal*, July 30, 1901, 4; *Sioux City Journal*, August 8, 1901, 9.
12. Sally H. Clarke, "Stress and Struggle inside International Harvester," *Enterprise and Society*, 22, no. 3 (September 2021): 665–67.
13. John Garraty, *Right-Hand Man: The life of George W. Perkins* (New York: Harper & Brothers, 1957), 94.
14. Cyrus McCormick, *Century of the Reaper* (Boston: Mifflin Co., 1931), 118.
15. *United States v. International Harvester Company*, Petitioners Exhibit 75, 135–36.
16. Luther Conant Jr., *The International Harvester Co.*(Department of Commerce and Labor, Bureau of Corporations, 1913), 22.
17. Ralph Baumheckel and Kent Borghoff, *International Harvester Farm Equipment* (American Society of Agricultural Engineers, 1997), 52.
18. William Z. Ripley, *Trusts, Pools and Corporations* (Boston: Ginn and Company, 1915), 634.
19. Wayne G. Broehl Jr., "Confronting International Harvester," in *John Deere's Company - Volume II* (Texas: Octane Press, 2023).
20. Broehl, "Confronting International Harvester."
21. Broehl, Appendix Exhibit 22, "Market Share, Deere and International Harvester, 1921 and 1929." in *John Deere's Company - Volume II.*
22. Clarke, "Stress and Struggle," 691.
23. *United States v. International Harvester Company*, 1, testimony of Henry B. Utley, 192.
24. *US v. International*, 1, testimony of Henry B. Utley, 193.
25. *The Implement Age* 23, no. 9 (May 1904): 27.
26. *The Implement Age* 23, no. 8 (April 1904): 24.
27. William Z. Ripley, *Trusts, Pools and Corporations* (Boston: Ginn and Company, 1915), 647.
28. *US v. International*, 1, testimony of Henry B. Utley, 195.
29. "Trust to Absorb Davenport Plant," *The Dispatch*, September 1, 1905, 5.
30. "Another Rumor," *The Implement Dealers' Bulletin*, September 1, 1905, 2.
31. "Trust Rumor is Denied," *The Fargo Forum and Daily Republican*, July 31, 1905, 10; "Tax Dodging and Bribery is Charge Brought by Woman," *The Inter Ocean from Chicago, Illinois*, June 21, 1905, 2.
32. "Says the Harvester Trust Got Millions in Railroad Rebates," *The Cairo Bulletin*, July 12, 1905, 1; "Charged with Conspiracy," *The York Dispatch*, July 21, 1905, 1.
33. C. H. McCormick Jr. to Harold McCormick, 1904, in Wisconsin Historical Society series 6z.
34. "Another Rumor," 2.
35. "Iowa Man Visiting Here," *Sterling Standard*, August 17, 1906, 2; *Sterling Standard*, April 5, 1904, 1.
36. R. B. Gray, *The Agricultural Tractor 1855-1950, Volume I* (American Society of Agricultural Engineers, 1954), 18.
37. Leffingwell, *Farmall: The Red Tractor*, 18.

38. John D. Culberson, *The Tractor Builders* (Iowa: Sunrise Hill Associates, 2001), 8.
39. Seyfarth, *Tractor History*, 26.
40. Sam Moore, "Ohio Manufacturing Company built early tractors," *Farm and Dairy*, May 22, 2014.
41. Seyfarth, *Tractor History*, 23.
42. Seyfarth, *Tractor History*, 28.
43. Baumheckel and Borghoff, *International Harvester Farm Equipment*, 90.
44. Seyfarth, *Tractor History*, 128.
45. Seyfarth, *Tractor History*, 29.
46. Seyfarth, *Tractor History*, 134.
47. Seyfarth, *Tractor History*, 137.
48. Seyfarth, *Tractor History*, 128.
49. Seyfarth, *Tractor History*, 118.
50. Seyfarth, *Tractor History*, 131.
51. Rick Mannen, *The Winnipeg Agricultural Motor Competitions 1908–1913* (self-pub., Canada, 2017).
52. Seyfarth, *Tractor History*, 29.
53. *The Commercial Motor*, August 8, 1909, 484.
54. *Harvester World*, August 1910.
55. Baumheckel and Borghoff, *International Harvester Farm Equipment*, 92.
56. Seyfarth, *Tractor History*, 119.
57. Seyfarth, *Tractor History*, 131.
58. Mannen, *The Winnipeg Agricultural Motor Competitions*.
59. Seyfarth, *Tractor History*, 132.
60. Seyfarth, *Tractor History*, 132.
61. Seyfarth, *Tractor History*, 128.
62. Original document reads "Mellapilla," which is likely "Melipilla" on modern Chilean maps.
63. Original document reads "Talcahano," which is likely "Talcahuano" on modern Chilean maps.
64. Seyfarth, *Tractor History*, 118.
65. Seyfarth, *Tractor History*, 121.
66. Seyfarth, *Tractor History*. 140.
67. Seyfarth, *Tractor History*.
68. Seyfarth, *Tractor History*, 128.
69. Seyfarth, *Tractor History*, 128.
70. Seyfarth, *Tractor History*, 132.
71. McCormick, *Century of the Reaper*, 173.
72. Sterling Evans, *Bound in Twine* (Texas A&M University Press, 2007), 162.
73. McCormick, *Century of the Reaper*, 174.
74. Guy Fay, *International Harvester Experimental and Prototype Tractors* (Minnesota: Motorbooks, 1997), 30.
75. Seyfarth, *Tractor History*, 126.
76. Naomi R. Lamoreaux, *The Great Merger Movement in American Business 1895–1904* (Massachusetts: Cambridge University Press, 1985), Table 1.1.
77. Ralph L. Nelson, *Merger Movements in American History, 1895–1956* (New Jersey: Princeton University Press, 1959), 39.
78. Lance Davis, "The Capital Markets and Industrial Concentration: The U.S. and U.K., a Comparative Study" *The Economic History Review*, 19, no. 2 (August 1966): 255–72.
79. Katie Wilson, "1912: When antitrust views collided in a presidential election," *Crosscut*, November 4, 2020.
80. C. H. Tavenner, "Harvester Trust Report Suppressed by Roosevelt," *The Steamboat Pilot*, May 15, 1912.
81. Evans, *Bound in Twine*, 98.
82. "Sees One Big Trust in Harvester-Steel; Stanley Declares He Has Such Evidence and Cites Federal Report Long Kept Secret," *The New York Times*, July 27, 1911.
83. "Harvester Trust to Reduce Size," *The New York Times*, August 5, 1918, 1, 9.
84. McCormick, *Century of the Reaper*, 187.
85. McCormick, *Century of the Reaper*, 182.
86. *International Harvester Industrial Power Activities*, Wisconsin Historical Society, 1940.
87. *International Harvester Industrial Power Activities*.

CHAPTER 3

1. Barton Currie, *The Tractor and its Influence Upon the Agricultural Implement Industry* (Pennsylvania: Curtis Publishing Company, 1916), 137–44.
2. Don Muhm, "At 92, A Father of the Farmall Tackles New Field," *Des Moines Sunday Register*, October 27, 1963.
3. A. C. Seyfarth, *Tractor History 1910–1934* (Wisconsin Historical Society, 2008), 116D.
4. Vernon Derry, *Plano, Illinois: Birthplace of The Harvester* (Illinois: Kelmscott Press, 1954).
5. Seyfarth, *Tractor History*, 17.
6. John F. Steward to C. S. Funk, September 9, 1910, in *Tractor History 1910–1934*, 87C.
7. Steward to Funk, in *Tractor History*, 87C.
8. *The Courier*, May 27, 1896, 3.
9. Steward to Funk, in *Tractor History*, 87C.
10. Seyfarth, *Tractor History*, 87A, 116C, 116I.
11. *Fortune*, August 1933, 28.
12. Pope Brock, *Charlatan* (New York: Crown Publishers, 2008), 72.
13. "Harold McCormick Quits as President of Harvester Firm," *New York Times*, June 3, 1922.
14. Seyfarth, *Tractor History*, 89
15. Ralph Baumheckel and Kent Borghoff, *International Harvester Farm Equipment* (American Society of Agricultural Engineers, 1997), 127.
16. "1943 ASME Gold Medal Winners," *Agricultural Engineering*, July 1943, in Bert Benjamin Archives.
17. E. C. Duffy, "Benjamin of Chicago: The Iowa Farm Boy Who Became a Famous Inventor," *The Sun and Agricultural Journal of S. A.*, May 1929, in Bert Benjamin Archives.
18. Bert R. Benjamin to R. R. Manatt, September 2, 1964, in Bert Benjamin Papers, Iowa State University.
19. Duffy. "Benjamin of Chicago."
20. C. W. Gray, "Notes on the Development of the Farmall Tractor," September 15, 1932, 3, in Wisconsin Historical Society, McCormick Archives, folder M95-063. Note that there are multiple copies of this document with mild variations. This is one of the few that has the Fordson reference.
21. Muhm, "A Father of the Farmall."
22. Bert R. Benjamin, "Proposed language for release," October 14, 1969, IH document, in Bert Benjamin Papers, Iowa State University.
23. Seyfarth, *Tractor History*, 93.
24. Seyfarth, *Tractor History*, 94.
25. Seyfarth, *Tractor History*, 6.
26. Muhm, "A Father of the Farmall."
27. Seyfarth, *Tractor History*, 98.
28. Muhm, "A Father of the Farmall."
29. Seyfarth, *Tractor History*, 100.
30. Seyfarth, *Tractor History*, 100.
31. Seyfarth, *Tractor History*, 102.
32. Seyfarth, *Tractor History*, 100.
33. Seyfarth, *Tractor History*, 105.
34. Seyfarth, *Tractor History*, 105.
35. Seyfarth, *Tractor History*, 111.
36. Seyfarth, *Tractor History*, 111.
37. Seyfarth, *Tractor History*, 111.

CHAPTER 4

1. Lawrence Svobida, *Farming the Dust Bowl: A First-Hand Account from Kansas* (University Press of Kansas, 1986), originally published as *Empire of Dust* (The Caxton Printers, 1940). Svobida farm story is sourced from this book.
2. International Harvester Company Annual Reports, 1930–1933, Wisconsin Historical Society.
3. Svobida, *Farming the Dust Bowl*, 164.
4. Svobida, *Farming the Dust Bowl*, 182.
5. International Harvester Annual Report, 1937, Wisconsin Historical Society.
6. Svobida, *Farming the Dust Bowl*, 233.
7. Lee Klancher, *John Deere Evolution* (Texas: Octane Press, 2021), 36–43.
8. Guy Fay, *International Harvester Experimental and Prototype Tractors* (Minnesota: Motorbooks, 1997), 75.
9. Oscar H. Will III and Todd Markle, *Farmall Regular and F-Series* (Minnesota: Voyageur Press, 2007), 50.
10. IH Company Decision, Wisconsin Historical Society
11. Will and Markle, *Farmall Regular and F-Series*, 60.
12. Will and Markle, *Farmall Regular and F-Series*, 61.
13. Will and Markle, *Farmall Regular and F-Series*, 34.
14. Will and Markle, *Farmall Regular and F-Series*, 67.
15. Ralph Baumheckel and Kent Borghoff, *International Harvester Farm Equipment* (American Society of Agricultural Engineers, 1997), 148.
16. Will and Markle, *Farmall Regular and F-Series*, 83.
17. Randy Leffingwell, *Farmall: The Red Tractor That Revolutionized Farming*, (Minnesota: Voyageur Press, 2007) 121–122.

CHAPTER 5

1. Daniel Feickert, *A Great Plains Farmer During the Great Depression* (University of Nebraska at Kearney, 2016), 7.
2. Feickert, *A Great Plains Farmer*, 17.
3. "The New Farmall A Takes a Bow," *Harvester World*, July 1938.
4. Timothy Egan, *The Worst Hard Time* (New York: Houghton Mifflin Publishing Company, 2006), 75.
5. "1940 International Harvester Annual Report," housed at Wisconsin Historical Society.
6. Raymond Loewy and Glenn Porter, *Never Leave Well Enough Alone* (John Hopkins University Press, 2002).
7. Paul Jodard, *Design Heroes: Raymond Loewy* (Harper Collins, 1994), 20.
8. Loewy and Porter, *Never Leave Well Enough Alone*.
9. Jodard, *Design Heroes*, 21.
10. Jodard, *Design Heroes*, 32.
11. "Modern Living," *Time*, October 31, 1949. The cover story for this issue was on Raymond Loewy.
12. R. MacDuff, Raymond Loewy Associates, to N. Hathaway, IH, October 15, 1957 memo, Raymond Loewy Collection, Library of Congress, Box 42, Folder 2. MacDuff informs Hathaway he was not able to find much about building prototypes.
13. Raymond Loewy to Gregg Montgomery, April 17, 1980, letter. Loewy states he was invited to work at IH by Fowler McCormick. Private correspondence not archived.
14. Edward Johnston, IH, and Raymond Loewy, June 25, 1936, contract, Raymond Loewy Collection, Library of Congress, Box 42, Folder 2.
15. Randy Leffingwell, *Farmall: The Red Tractor that Revolutionized Farming* (Minnesota: Voyageur Press, 2007), 149.
16. Raymond Loewy Associates to P. W. Holder, IHC, August 2, 1946, contract letter, Raymond Loewy Collection, Library of Congress, Box 42, Folder 2.
17. Fred Toerge, Contact Report for January 6, 1967, Raymond Loewy Collection, Library of Congress, Box 64, Folder 13, "Business Promotion 1967–77" file.
18. Meeting itinerary for January 6, 1967, Library of Congress, Box 170, Folder 14, Raymond Loewy Project file.
19. Fred Toerge, Contact Report for January 6, 1967, Raymond Loewy Collection, Library of Congress, Box 64, Folder 13, "Business Promotion 1967–77" file.
20. *Time*, October 31, 1949.
21. *Time*, October 31, 1949.
22. Interview transcript, Gregg Montgomery, recorded February 1, 2023.
23. Raymond Loewy to Gregg Montgomery, April 17, 1980. Private correspondence not archived.
24. "Farmall MCH?" Yesterday's Tractor Co., forum discussion.
25. IH parts catalog, Farmall M and Farmall Super M.
26. "A Big Man for Small Farms: Harry Merritt helped make tractors possible for the little guy," *OEM Off-Highway News*, May 1, 2005.
27. Wayne G. Broehl Jr., Exhibit 11-5, "Long-Line Farm Equipment Sales 1929–1936" in *John Deere's Company Volume II* (Texas: Octane Press, 2023).
28. Broehl Jr., Appendix: Exhibit 29, "Farm Equipment Sales in North America 1970–1980" in *John Deere's Company Volume II*.
29. Leffingwell, *Farmall: The Red Tractor*, 132.

30. Guy Fay, *International Harvester Experimental and Prototype Tractors* (Minnesota: Motorbooks, 1997), 101.
31. Fay, *International Harvester Experimental*, 106–7.
32. "The New Farmall A Takes a Bow."
33. Jessica Traynor, "Harry Ferguson, the 'Mad Mechanic' who invented the modern tractor," *Irish Times*, February 26, 2019.
34. Ed Stembridge, "Ag History: Ford N Series Tractors and the Handshake that Forever Changed Farming," *Curbside Classic*, February 21, 2021.
35. Stembridge, "Ag History."

CHAPTER 5 (CAPTIONS)

i. 2004255_0019, Box 1, Folder "Automobile Cartoons," Raymond Loewy collection of photographs and audiovisual materials (Accession 2004.255), Audiovisual Collections and Digital Initiatives Department, Hagley Museum and Library, Wilmington, DE 19807.
ii. 20111207_Studebaker1939, Series I. Automobile Makes, Box 1-827, Z. Taylor Vinson collection of transportation ephemera (Accession 20100108.ZTV), Published Collections Department, Hagley Museum and Library, Wilmington, DE 19807.
iii. 2004255_0155, Box 4, Folder "Homes - La Cense, France," Raymond Loewy collection of photographs and audiovisual materials (Accession 2004.255), Audiovisual Collections and Digital Initiatives Department, Hagley Museum and Library, Wilmington, DE 19807. Raymond Loewy™ by Good Design, LLC, RaymondLoewy.com.
iv. Randy Leffingwell, *Farmall: The Red Tractor that Revolutionized Farming* (Minnesota: Voyageur Press, 2007), 137.

CHAPTER 6 (BIBLIOGRAPHY)

Berndt, Thomas. *Standard Catalog of US Military Vehicles 1940–1965*. Wisconsin: Krause Publications, 1995.
Crismon, Frederick W. *International Trucks*. 3rd ed. Victory WW2 Publishing Ltd., 2007.
Crismon, Frederick W. *US Military Wheeled Vehicles*. Minnesota: Motorbooks International, 1994.
Jackson, David D. "International Harvester in World War Two." The American Auto Industry in World War Two. Updated March 2, 2023.
McCormick-International Harvester Collection, Wisconsin Historical Society.
"Tank Encyclopedia," The Online Tank Museum.
US Army. *Modern Ordnance Materiel*. New Jersey: Raritan Arsenal, 1943.
US Army, *TM9-2800*, Washington: United States Government Printing Office, September 1943.
Vanderveen, Bart H. *The Observer's Army Vehicles Directory to 1940*. United Kingdom: Frederick Warne & Company, 1974.
Vanderveen, Bart H. *The Observer's Fighting Vehicles Directory, World War II*. United Kingdom: Frederick Warne & Company, 1969.

CHAPTER 7

1. John Hildebrand, *Mapping the Farm: The Chronicle of a Family* (New York: Vintage / Random House, 1995).
2. Lowry Nelson, *American Farm Life* (Massachusetts: Harvard University Press, 1954), 36.
3. Nelson, Table 7, "Percentage of United States farms reporting motor vehicles, 1920–1950," in *American Farm Life*, 33.
4. USDA Census of Agriculture, "Farms and Farm Property," in *Volume 3: General Report of 1940*, housed at Cornell University Library.
5. USDA Census of Agriculture, 1950, housed at Cornell University Library.
6. John L. McCaffrey, IH, to Committee on Agriculture, US House of Representatives, February 13, 1948, housed at University of Wisconsin Archives.
7. Nelson, Figure 1, "Changes in Farm Size, by Regions, 1920 to 1950," in *American Farm Life*, 20.
8. Bruce L. Gardner, Figure 1.1, "U.S. land in farms," in *American Agriculture in the Twentieth Century: How It Flourished and What It Cost* (Massachusetts: Harvard University Press, 2022), 4.
9. Nelson, Table 6, "Changes in land tenure and average acreage per farm, 1930–1950," in *American Farm Life*, 26.
10. Wayne G. Broehl Jr., Appendix: Exhibit 38, "Deere Profitability Margins" in *John Deere's Company Volume II*. See also Exhibit 12 and 26.
11. *Harvester World* magazine, Oct-Nov-Dec 1945, 4, housed at Wisconsin Historical Society.
12. Ralph Baumheckel and Kent Borghoff, *International Harvester Farm Equipment* (American Society of Agricultural Engineers, 1997), 171.
13. *Harvester World* magazine, Oct-Nov-Dec 1945.
14. *Harvester World* magazine, Oct-Nov-Dec 1945.
15. *International Harvester News Releases: 1947*, June 11, 1947, housed at Wisconsin Historical Society.
16. Raymond Loewy, *Industrial Design* (New York: The Overlook Press, 1979), 126.
17. "IH Annual Report," 1944, housed at Wisconsin Historical Society.
18. Raymond Loewy and IH contract, dated August 20, 1943, signed by J. M. McCaffrey, Second Vice President, housed at Raymond Loewy Collection in Library of Congress.
19. "International Harvester Company Program for General Line Branches," Raymond Loewy Associates, June 4, 1938, Raymond Loewy Collection, Library of Congress, Box 149, Folder 5.
20. *International Harvester News Releases: 1947*, July 30, 1937, housed at Wisconsin Historical Society.
21. *International Harvester News Releases: 1947*, August 28, 1937, housed at Wisconsin Historical Society.
22. "International Harvester Holds Big Power Show," *International Harvester News Releases: 1947*, housed at Wisconsin Historical Society.
23. "International Harvester Holds Big Power Show."
24. "International Harvester Holds Big Power Show."
25. Oscar H. Will, *Payline* (Minnesota: Motorbooks, 2006), 25.
26. Will, *Payline*, 26.
27. Barbara Marsh, *A Corporate Tragedy* (New York: Doubleday & Company, 1985), 78.
28. Baumheckel and Borghoff, *International Harvester Farm Equipment*, 174–76.
29. Guy Fay and Andy Kraushaar, *Farmall Letter Series Tractors* (Minnesota: MBI Publishing, 1998), 45–49.
30. Brian Wayne Wells, "M&W Company of Anchor, Illinois." *Belt Pulley* magazine, November/December 1997.
31. Dale Bielfeldt, interview by Max Armstrong, September 10, 2022.
32. Bill Burnham, interview by Lee Klancher, May 25, 2023.
33. Bill Burnham, interview by Lee Klancher, May 25, 2023.
34. Bob Meiners, interview by Max Armstrong, September 10, 2022.
35. Bill Burnham, interview by Lee Klancher, May 25, 2023.
36. *Bloomington Paragraph*, article, March 19, 1949.
37. Wells, "M&W Company."
38. Dale Bielfeldt, interview by Max Armstrong, September 10, 2022.
39. Ron Meiners, interview by Max Armstrong, September 10, 2022.
40. Bill Burnham, interview by Lee Klancher, May 25, 2023.
41. Ron Meiners interview.
42. Bill Burnham, interview by Lee Klancher, May 25, 2023.
43. "Elmo R. Meiners of Fairbury, Illinois, |1914–2009| Obituary," Duffy-Pils Memorial Home.
44. "Selling," *IH Today*, July-August 1950, 21.
45."The Mid-Century Way," *Harvester World*, February-March 1950.
46. "Selling," *IH Today*, 21.
47. Fay and Kraushaar, *Farmall Letter Series Tractors*, 35–44.
48. Baumheckel and Borghoff, *International Harvester Farm Equipment*, 160–61.
49. Guy Fay, *Original Farmall Letter Series*, 103.
50. Baumheckel and Borghoff, *International Harvester Farm Equipment*, 160–61.
51. Guy Fay, *Original Farmall Letter Series*, 77.
52. "What's a Stage II Super H?, Red Power Forum, Serial Number breaks from former employee, Dr. Evil, January 26, 2005. https://www.redpowermagazine.com/forums/topic/950-whats-a-stage-ii-super-h-mean/
53. Stage I vs Stage II Super H, January 14, 2008, Yesterday's Tractors online discussion board. https://www.yesterdaystractors.com/cgi-bin/viewit.cgi?bd=farmall&th=637291
54. Guy Fay, *International Harvester Tractor Data Book* (Minesota: Motorbooks, 1997), 134.
55. Guy Fay, *Original Farmall Letter Series*, 77.
56. Guy Fay, *International Harvester Experimental and Prototype Tractors* (Minnesota: Motorbooks, 1997), 111.
57. "Under the Big Top," *Harvester World*, July/August 1953, 16.
58. "The Show on Machinery Hill," *Harvester World*, October 1953, 13.
59. Gordon L. Hershman, interview by Lee Klancher, circa 1995.
60. Robert Oliver, interview by Lee Klancher, circa 1995.

CHAPTER 8

1. Bill Burnham, interview by Lee Klancher, May 25, 2023. This interview is referenced throughout the story about Bill's father, William.
2. "Current Population Reports," March 25, 1952, Series P-60, No. 9. According to Census Bureau data, average US farm income in 1950 was $2,000. Average US income at that time was $3,300.
3. "Figures from IH Annual Reports, 1954–1958," housed at University of Wisconsin Archives.
4. Bruce L. Gardner, Figure 2.3a, "Labor Hours per unit of crops," in *American Agriculture in the Twentieth Century* (Massachusetts: Harvard University Press, 2002). 16.
5. Gardner, Labor Figure 2.3b, "Labor Hours per unit livestock output," in *American Agriculture*, 17.
6. M&W advertisement in *The Pantagraph*, September 8, 1956.
7. "Figures from IH Annual Reports, 1909–1959," housed at University of Wisconsin Archives.
8. *IH Press Releases 1956*, January 20, 1956, page 6. Wisconsin Historical Society.
9. Guy Fay, *Original Farmall Hundred Series 1954–1958* (Minnesota: Motorbooks International, 2003), 30.
10. Fay, *Original Farmall Hundred Series*, 43.
11. Ralph Baumheckel and Kent Borghoff, *International Harvester Farm Equipment* (American Society of Agricultural Engineers, 1997), 186.
12. Fay, *Original Farmall Hundred Series*, 77.
13. Fay, *Original Farmall Hundred Series*, 108.
14. Bob Meiners, interview by Max Armstrong, September 10, 2022.
15. *IH News Release*, June 22, 1954, Wisconsin Historical Society Archives.
16. Fay, *Original Farmall Hundred Series*, 108.
17. IH Brochure, "The Electrall Motor," references powering the McCormick No. 55 Baler.
18. IH Brochure, "Electrall Tractor Generator."
19. Fay, *Original Farmall Hundred Series*, 108.
20. Fay, "Model 600 monthly production figures," in *Original Farmall Hundred Series*, 136.

CHAPTER 8 (CAPTION)

i. Images from Wabash County Museum identify the date and the farm where the contest was held.

There are a number of people I'd like to thank for their help in making this work come to life.

Author and researcher Guy Fay worked for years at the McCormick Archives and has uncovered so many of the significant pieces of research material on IH tractors to surface in the past decade. When you read a book about IH, Guy Fay is living and breathing on the pages. Guy is the one who discovered the Keystone tractor and John F. Steward connections in the book, among other things.

Todd Markle is another IH writer who has done good work, and his help with the F-series tractors was also terrific. Jim Allen's chapter on IH's war production and other contributions are equally stellar, and Kenneth Updike provided several interesting takes on IH technology.

The book reviewers who read the manuscript and helped refine the work were invaluable. Howard Raymond, Chuck Johnson, Todd Markle, Tracey Meiners, and Jim Becker were the standouts on this front, offering great insight that elevated the work.

Max Armstrong pointed me to Bill Burnham and shared his interviews with Bill. Max is a great supporter and friend of these books—it's always a pleasure to work with him. Bill Burnham is a historian in his own right, and he shared information and his tractors for three days at his place.

To all the collectors who took time to clean machines, dig out equipment, and patiently stood by while I scouted spots and waited for the light to get right, a huge thanks. The photography in this book was made over the past thirty years, so there are too many to name, but John Kinzenbaw (and Garry Roberts), J. W. Goodwin (and Kevin Blackburn), Ryan Peters, and Jerry Kuster were the recent standouts. A big shout-out to the folks at the Half Century Show for helping with the studio images made there in 2019, particularly the resourceful and unflaggingly helpful Russell Buhr and John Frederickson.

Sarah Pickett at Case IH offered good insight and support in producing the work—she is the most knowledgeable licensing person on the planet!

I'd like to thank Sally J. Jacobs, for her help pointing me to online resources to search the McCormick Archives housed at the Wisconsin Historical Society, and all the helpful people at Wisconsin Historical Society who came through to get our images delivered.

Designer Tom Heffron powered through creating this work, building pages and also catching all kinds of details. He's like an editor and designer rolled into one!

I also owe thanks to the editor, research assistant, and all-around ace Faith Garcia. She shepherded the book as copy and production editor and also traveled to the Library of Congress to find the terrific new materials uncovered from Raymond Loewy. Her hard work, attention to detail, and critical eye shaped this work.

Last but not least, thanks to my wife, Joan, who adds plenty of critical feedback but, more importantly, without whom none of this would matter much.

HARD AT WORK

These Farmalls are shown working the fields of Wabash Farms in Indiana in a photo dated May 1938. *Arthur Rothstein, Library of Congress LC-DIG-fsa-8a09722*

MODEL INDEX

IH and Farmall Tractors